THE BOOK OF

HOPE

ALSO BY JANE GOODALL

My Friends the Wild Chimpanzees

Innocent Killers

In the Shadow of Man

The Chimpanzees of Gombe: Patterns of Behavior

Through a Window: My Thirty Years with the Chimpanzees of Gombe

Visions of Caliban: On Chimpanzees and People

Brutal Kinship

Reason for Hope: A Spiritual Journey

Africa in My Blood: An Autobiography in Letters: The Early Years

Beyond Innocence: An Autobiography in Letters: The Later Years

The Ten Trusts: What We Must Do to Care for the Animals We Love

Harvest for Hope: A Guide to Mindful Eating (with Gary McAvoy and Gail Hudson)

Hope for Animals and Their World: How Endangered Species Are Being Rescued from the Brink (with Thane Maynard and Gail Hudson)

Seeds of Hope: Wisdom and Wonder from the World of Plants (with Gail Hudson and Michael Pollan)

THE GLOBAL ICONS SERIES

The Book of Joy: Lasting Happiness in a Changing World by His Holiness the Dalai Lama and Archbishop Desmond Tutu with Douglas Abrams

The Book of Hope: A Survival Guide for an Endangered Planet by Jane Goodall and Douglas Abrams

THE BOOK OF

HOPE

A SURVIVAL GUIDE FOR
AN ENDANGERED PLANET

JANE GOODALL

AND

DOUGLAS ABRAMS

with GAIL HUDSON

VIKING
an imprint of
PENGUIN BOOKS

VIKING

UK | USA | Canada | Ireland | Australia
India | New Zealand | South Africa

Viking is part of the Penguin Random House group of companies
whose addresses can be found at global.penguinrandomhouse.com.

First published in the United States of America by Celadon Books 2021
First published in Great Britain by Viking 2021
001

Printed and bound in Great Britain by Clays Ltd, Elcograf S.p.A.

The authorized representative in the EEA is Penguin Random House Ireland,
Morrison Chambers, 32 Nassau Street, Dublin D02 YH68

A CIP catalogue record for this book is available from the British Library

ISBN: 978–0–241–47857–8

www.greenpenguin.co.uk

Penguin Random House is committed to a
sustainable future for our business, our readers
and our planet. This book is made from Forest
Stewardship Council® certified paper.

To Mum, Rusty, Louis Leakey, and David Greybeard

—JANE GOODALL

To my parents, and to Hassan Edward Carroll

and all who struggle to find hope

—DOUG ABRAMS

Contents

(JANE GOODALL INSTITUTE/BILL WALLAUER)

An Invitation to Hope

We are going through dark times.

There is armed conflict in many parts of the world, racial and religious discrimination, hate crimes, terrorist attacks, a political swing to the far right fueling demonstrations and protests that, all too often, become violent. The gap between the haves and have-nots is widening and fomenting anger and unrest. Democracy is under attack in many countries. On top of all that, the COVID-19 pandemic has caused so much suffering and death, loss of jobs, and economic chaos around the world. And the climate crisis, temporarily pushed into the background, is an even greater threat to our future—indeed, to all life on Earth as we know it.

Climate change is not something that might affect us in the future—it is affecting us now with changing weather patterns around the globe: melting ice; rising sea levels; and catastrophically powerful hurricanes, tornadoes, and typhoons. There is worse flooding, longer droughts, and devastating fires that are breaking out around the globe. For the first time, fires have even been recorded in the Arctic Circle.

"Jane is almost ninety years old," you may be thinking. "If she is aware of what is going on in the world, how can she still be writing about hope? She is probably giving in to wishful thinking. She is not facing up to the facts."

I am facing up to the facts. And on many days I admit that I feel depressed, days when it seems that the efforts, the struggles, and the sacrifices of so many people fighting for social and environmental justice, fighting prejudice and racism and greed, are fighting a losing battle. The forces raging around us—greed, corruption, hatred, blind prejudice—are ones we might be foolish to think we can overcome. It's understandable that there are days we feel we are doomed to sit back and watch the world end "not with a bang but a whimper" (T. S. Eliot). Over the last eight decades I have been no stranger to disasters such as 9/11, school shootings, suicide bombings, and so on, and the despair that some of these terrible events can elicit. I grew up during World War II, when the world risked being overrun by Hitler and the Nazis. I lived through the Cold War arms race, when the world was threatened by a thermonuclear holocaust, and the horrors of the many conflicts that have condemned millions to torture and death around the globe. Like all people who live long enough, I have been through many dark periods and seen so much suffering.

But each time I become depressed I think of all the amazing stories of the courage, steadfastness, and determination of those who are fighting the "forces of evil." For, yes, I do believe there is evil amongst us. But how much more powerful and inspirational are the voices of those who stand up against it. And even when they lose their lives, their voices still resonate long after they are gone, giving us inspiration and hope—hope in the ultimate goodness of this strange, conflicted human animal that evolved from an apelike creature some six million years ago.

Ever since I began traveling around the world in 1986 to raise awareness about the harm we humans have created, socially and environmentally, I have met so many people who have told me they have lost hope for the future. Young people especially have been angry, depressed, or just apathetic because, they've told me, we have compromised their future and they feel there is nothing they can do about it. But while it is true that we have not just compromised but stolen their future as we have relentlessly plundered the finite resources of our planet with no concern for future generations, I do not believe it is too late to do something to put things right.

Probably the question I am asked more often than any other is: Do you honestly believe there is hope for our world? For the future of our children and grandchildren?

And I am able to answer truthfully, yes. I believe we still have a window of time during which we can start healing the harm we have inflicted on the planet—but that window is closing. If we care about the future of our children and theirs, if we care about the health of the natural world, we must get together and take action. Now—before it is too late.

What is this "hope" that I still believe in, that keeps me motivated to carry on, fighting the good fight? What do I really mean by "hope"?

Hope is often misunderstood. People tend to think that it is simply passive wishful thinking: I hope something will happen but I'm not going to do anything about it. This is indeed the opposite of real hope, which requires action and engagement. Many people understand the dire state of the planet—but do nothing about it because they feel helpless and hopeless. That is why this book is important, as it will, I hope (!), help people realize that their actions, however small they may seem, will truly make a difference. The cumulative effect of thousands of ethical actions can help to save and improve our world for future generations. And why would you bother to take action if you did not truly hope that it would make a difference?

My reasons for hope in these dark times will become clear in this book, but for now let me say that without hope, all is lost. It is a crucial survival trait that has sustained our species from the time of our Stone Age ancestors. Certainly, my own improbable journey would have been impossible had I lacked hope.

All of this and more I discussed with my coauthor, Doug Abrams, throughout the pages of this little book. Doug proposed the book as a dialogue similar to that of *The Book of Joy*, which he had written with the Dalai Lama and Archbishop Desmond Tutu. In the chapters that follow, Doug serves as the narrator, sharing the dialogues that took place between us in Africa and Europe. With Doug's help, I am now able to share with you what I have learned about hope throughout my long life and study of the natural world.

Hope is contagious. Your actions will inspire others. It is my sincere desire that this book will help you find solace in a time of anguish, direction in a time of uncertainty, courage in a time of fear.

We invite you to join us on this journey toward hope.

Jane Goodall, Ph.D., DBE, UN Messenger of Peace

Reaching across the nonexistent barrier once thought to divide us from the rest of the animal kingdom. (JANE GOODALL INSTI- TUTE/HUGO VAN LAWICK)

I

What Is Hope?

Whisky and Swahili Bean Sauce

It was the night before we were to begin our dialogues. I was nervous—because the stakes were high. The world seemed to need hope more than ever, and in the months since reaching out to Jane to ask if she wanted to share her reasons for hope in a new book, the subject of hope had been uppermost in my thoughts. What is it? Why do we have it? Is hope real? Can hope be cultivated? Is there really hope for our species? I knew my role was to ask the questions we all wrestle with as we experience adversity and even, at times, despair.

Jane is a global hero who has traveled the world for decades as a messenger of hope, and I was eager to understand her confidence in the future. Equally, I wanted to know how she had sustained hope during her own challenging and pioneering life.

As I was eagerly and anxiously preparing my questions, the phone rang.

"Would you like to come around for dinner with the family?" Jane asked. I had just landed in Dar es Salaam, and I told her I would be delighted to join her and meet her family. It would be a chance not just to meet the icon but to see her as mother and grandmother; to break bread; and, as I suspected, to sip whisky.

Finding Jane's house is not easy, as there is no real street address. It is down a number of dirt roads and next to the large compound of

Julius Nyerere, the first president of Tanzania. I was afraid I might be late as the taxi tried unsuccessfully to find the right entrance in the tree-covered neighborhood. The red sun was descending quickly and there were no streetlights to guide us.

When we finally found the house, Jane greeted me at the door with a warm smile and wide, penetrating eyes. Her gray hair was pulled back in a ponytail, and she wore a green button-down shirt and khaki pants, which looked a little like the uniform of a park ranger. On her shirt was a logo for the Jane Goodall Institute (JGI) with the symbols of the organization: a profile of Jane, a chimpanzee knuckling on all fours, a leaf for the environment, and a hand for the humans that she has come to realize need protection along with the chimps.

Jane is eighty-six, but inexplicably she doesn't seem to have aged very much since she first went to Gombe and graced the cover of *National Geographic*. I wondered if there is something about hope and purpose that keeps one endlessly young.

But what stands out most is Jane's will. It shines from her hazel eyes like a force of nature. It is the same will that first moved her halfway around the globe to study animals in Africa and has kept her traveling for the last thirty years. Before the pandemic, she was spending more than three hundred days a year lecturing about the risks of environmental destruction and habitat loss. Finally, the world is starting to listen.

I knew that Jane liked her evening whisky and had brought her a bottle of her favorite, Green Label Johnnie Walker. She graciously accepted it—but later she told me I should have bought the cheaper one, Red Label, and donated the extra money to her environmental organization, the Jane Goodall Institute.

In the kitchen, Maria, her daughter-in-law, had prepared a Tanzanian vegetarian meal. There was coconut rice served with a creamy Swahili bean sauce; lentils and peas with a hint of ground peanuts, curry, and coriander; and sautéed spinach. Jane says she cares nothing about food, but I can't say the same and my mouth was already beginning to water.

She placed my little gift on the counter next to a giant, four-and-a-half-liter bottle of Famous Grouse whisky. Jane's adult grandchildren had gotten it for her as a surprise, and they explained that it was so much cheaper to buy in bulk and would surely last for the time she would be with them. Her grandchildren live in the house in Dar es Salaam where Jane moved when she married her second husband, though in those days most of her time was still spent in Gombe. Now Jane spends time in the house only during her short twice-a-year visits to Tanzania and only for a few days at a time, as she also goes back to Gombe and other towns in Tanzania.

For her, an evening tot of whisky is a nightly ritual and an opportunity to relax and, when possible, toast with friends.

"It all started," she explained, "because Mum and I always shared a 'wee dram' every evening when I was at home. So we went on raising a glass to each other at 7 p.m. wherever I was in the world." She has also found that when her voice gets really tired from too many interviews and lectures, a small sip of whisky tightens the vocal cords and enables her to get through a lecture. "And," said Jane, "four opera singers and one popular rock singer have told me that this works for them, too."

I sat next to Jane at the outdoor table on the veranda as she and her family laughed and told stories. The thick bougainvillea surrounding us almost felt like a forest canopy in the candlelight.

With my family in Dar es Salaam. Left to right: grandson Merlin; his half brother Kiki, son of Maria; my grandson Nick, half brother to Merlin; granddaughter Angel; and my son Grub. (JANE GOODALL INSTITUTE/COURTESY OF THE GOODALL FAMILY)

Merlin, her eldest grandson, was twenty-five years old. Years earlier, when he was eighteen, after a wild night with friends he had dived into an empty swimming pool. He was left with a broken neck, and the injury had caused him to change his life, to give up partying, and, like his sister Angel, follow his grandmother into conservation work. Jane, the understated matriarch, sat at the head of the table, her pride clearly evident.

Jane put mosquito repellent on her ankles and we joked that the mosquitos were not vegetarians. "Only the female sucks blood," Jane pointed out. "The males just live off nectar." Through the eyes of the naturalist, the bloodsucking mosquitoes were simply mothers who were trying to get a blood meal to feed their offspring. That didn't quite change my dislike of these historic foes of humanity, however.

Angel is working with our Roots & Shoots program and Merlin is helping to develop an education center in an ancient remnant forest near Dar es Salaam. (K 15 PHOTOS/FEMINA HIP)

As the conversation and family stories paused, I wanted to ask Jane the questions that had been absorbing me ever since we first decided to collaborate on a book about hope.

As a born-and-raised and somewhat skeptical New Yorker, I had to admit that I was suspicious of hope. It seemed like a weak response, a passive acceptance—"let's hope for the best." It seemed like a panacea or a fantasy. A willful denial or blind faith to cling to despite the facts and the grim reality of life. I was afraid of having false hope, that misguided imposter. Even cynicism felt safer in some ways than taking the risk of hope. Certainly, fear and anger seemed like more useful responses, ready to sound the alarm, especially during times of crisis like this.

I also wanted to know what the difference was between hope and optimism, whether Jane had ever lost hope, and how we keep hope in dark times. But these questions would need to wait until the next morning, as it was getting late and the dinner party was breaking up.

Is Hope Real?

When I returned the next day—a little less nervous—to begin our conversation about hope, Jane and I sat on her veranda in old, sturdy wooden folding chairs with green canvas seats and backs. We looked out at the backyard so filled with trees that it was almost impossible to see the Indian Ocean just beyond. A chorus of tropical birds sang, screeched, cackled, and called. Two rescue dogs came to curl up at Jane's feet, and a cat meowed through a screen, insistent about contributing to the conversation. Jane seemed a little like a modern-day Saint Francis of Assisi, surrounded by and protecting all the animals.

"What is hope?" I began. "How do *you* define it?"

"Hope," Jane said, "is what enables us to keep going in the face of adversity. It is what we desire to happen, but we must be prepared to work hard to make it so." Jane grinned. "Like hoping this will be a good book. But it won't be if we don't bloody work at it."

I smiled. "Yes, that is definitely one of my hopes, too. You said that hope is what we desire to happen, but we need to be prepared to work hard. So does hope require action?"

"I don't think all hope requires action, because sometimes you can't take action. If you're in a cell in a prison where you've been thrown for no good reason, you can't take action, but you can still hope to get out. I've been communicating with a group of conservationists who have been tried and given long sentences for putting up camera traps to record the presence of wildlife. They're living in hope for the day they're released through the actions of others, but they can't actually take action themselves."

It sounded like action and agency were important for generating hope, but that hope could survive even in a prison cell. A black cat with a white chest strolled out of the house and onto the balcony and jumped in Jane's lap, curling up comfortably, his paws tucked under him.

"I'm wondering if animals have hope."

Jane smiled. "Well, when Bugs here," she said, petting the cat, "was sitting inside all that time, I suspect he was 'hoping' that eventually he would be let out. When he wants food, he gives plaintive meows and rubs against my legs with arched back and waving tail, as this usually produces the desired effect. I'm sure when he does that he's hoping he will be fed. Think of your dog waiting in the window for you to come home. That's clearly some form of hope. Chimps will often throw a tantrum when they don't get what they want. That is some form of frustrated hope."

It seemed like hope was not uniquely human, but I knew we'd return to what made hope unique in the human mind. For now, I wanted to understand how hope was different from another term with which it is often confused. "Many of the world's religious traditions talk about hope in the same breath as faith," I said. "Are hope and faith the same?"

"Hope and faith are very different, aren't they," Jane said, more as a statement than a question. "Faith is when you actually believe there is an intellectual power behind the universe, which can be translated into God or Allah or something like that. You believe in God, the Creator. You believe in life after death or some other doctrine. That's faith. We can *believe* that these things are true, but we can't *know*. But we can know the direction we want to go and we can *hope* that it is the right direction. Hope is more humble than faith, since no one can know the future."

"You were saying that hope requires us to work hard to make what we want to happen actually happen."

"Well, in certain contexts it is essential. Take this dire environmental nightmare we are living in today. We certainly hope that it is not too late to turn things around—but we know that this change will not happen unless we take action."

"So by being active, you become more hopeful?"

"Well, you have it both ways. You won't be active unless you hope that your action is going to do some good. So you need hope to get you going, but then by taking action, you generate more hope. It's a circular thing."

"So what actually is hope—an emotion?"

"No, it's not an emotion."

"So what is it?"

"It's an aspect of our survival."

"Is it a survival skill?"

"It's not a skill. It's something more innate, more profound. It's almost a gift. Come on, think of another word."

"'Tool'? 'Resource'? 'Power'?"

"'Power' would do. 'Power'—'tool.' Something like that. Not a power tool!"

I laughed at Jane's joke. "Not a drill?"

"No, not an electric drill," Jane said, laughing, too.

"A survival mechanism . . . ?"

"Better, but less mechanical. A survival . . ." Jane paused, trying to come up with the right word.

"Impulse? Instinct?" I offered.

"Actually, it's a survival trait," she finally concluded. "That's what it is. It is a human survival trait and without it we perish."

If it was a survival trait, I wondered why some people had more of it than others, if it could be developed during particularly stressful times, and whether she had ever lost it.

Have You Ever Lost Hope?

Jane has a rare blend of qualities   a scientist's unflinching willingness to face the hard facts and a seeker's desire to understand the most profound questions of human life.

"As a scientist, you—" I began.

"I consider myself a naturalist," she corrected.

"What's the difference?" I had always assumed a naturalist was simply a scientist who went out into the field.

"The naturalist," Jane said, "looks for the wonder of nature—she listens to the voice of nature and learns from nature as she tries to understand it. Whereas a scientist is more focused on facts and the desire to quantify. For a scientist, the question is, 'Why is this adaptive? How does it contribute to the survival of the species?'

"As a naturalist, you need to have empathy and intuition—and love. You've got to be prepared to look at a murmuration of starlings and be filled with awe at the amazing agility of these birds. How do they fly in a flock of several thousand without touching at all, and yet have such close formations, and swoop and turn together almost as one? And why do they do it—for fun? For joy?" Jane looked up at the imagined starlings, and her hands danced as if they were a flock of birds rippling through the sky.

I could suddenly see Jane as a young naturalist full of awe and wonder. When the rain started to pour down loudly, pausing our conversation, it was not hard to imagine her back in those early days when her own hopes and dreams seemed so distant and so difficult to realize.

When the rain quieted, we resumed our conversation. I asked Jane what she remembered about her first journey to Africa. She closed her eyes. "It was like a fairy tale," she said. "There were no planes flying back and forth in those days—it was 1957—so I went by boat, the *Kenya Castle*. It should have taken about two weeks but ended up being about a month because there was a war between England and Egypt, so the Suez Canal was closed. We had to go right around the whole African continent, down to Cape Town and up the coast to Mombasa. A magical voyage."

Jane was in pursuit of her dream to study animals in the wild, a dream that had been born as a child reading Doctor Doolittle and Tarzan stories. "Tarzan clearly married the wrong Jane," she joked. The improbability of Jane's life has inspired many across the globe. At the time, women did not travel halfway around the world to go into the jungle to live with and write books about wild animals. As Jane said, "Even men were not doing that!"

I asked her to tell me more about those early days.

"I did very well at school," she said, "but when I graduated at eighteen there was no money for university. I had to get a job, so I did a secretarial course. Boring stuff. But Mum had told me that I would have to work hard and take advantage of opportunities, and not to give up.

"Mum always used to say, 'If you're going to do a thing, do it properly.' I think that's been a cornerstone of my life. You don't want to do it, you want to get it over with, but if you're going to do it at all, then put the best you have into it."

Jane's opportunity came when a school friend invited her to visit her family's farm in Kenya. And it was during that visit that she heard about Dr. Louis Leakey, the famed paleoanthropologist, who had spent his life searching for the fossilized remains of our earliest human ancestors in Africa. At the time he was curator of the Coryndon Museum (now called the Nairobi National Museum).

"Someone told me that if I was interested in animals, I should meet Leakey," Jane said. "So I made an appointment to see him. I think he was impressed by how much I knew about African animals—I'd read everything I could about them. And guess what—two days before I met him his secretary had suddenly left, and he needed one. So you see, that boring old secretarial training paid off after all!"

She was invited to join Leakey; his wife, Mary; and Gillian, another young Englishwoman, on their annual dig at Olduvai Gorge in Tanzania, searching for early human remains.

"Toward the end of the three months, Louis began talking about a group of chimpanzees living in the forests along the eastern shore of Lake Tanganyika in Tanzania, which at the time was called Tanganyika and still under British colonial rule. He told me that the chimpanzee habitat was remote and rugged and that there would be dangerous animals—and that chimpanzees themselves were four times stronger than humans. Oh, how I longed to undertake an adventure like the one

With Dr. Louis S. B. Leakey—the man who made my dream come true.
(JANE GOODALL INSTITUTE/JOAN TRAVIS)

Leakey was envisioning. He said he was looking for someone with an open mind, with a passion for learning, a love of animals, and endless patience."

Leakey believed that an understanding of how our closest relatives behaved in the wild might shed light on human evolution. He wanted someone to do this study because, while you can tell a great deal about what a creature looked like from the skeleton and about its diet from tooth wear, *behavior* does not fossilize. He believed there was a common ancestor, an apelike humanlike creature, some six million years ago. He reasoned that if modern chimpanzees (with whom we share almost 99 percent of the composition of our DNA) showed behavior similar (or identical) to that of modern humans, it might have been present in that common ancestor and been part of our repertoire throughout our separate evolutionary pathways. And this, he thought, would enable him to better guess the behavior of our Stone Age ancestors.

"I had no idea he was thinking of me," Jane said, "and I could hardly believe it when he asked if I was prepared to undertake this task!" Jane smiled as she recalled her mentor. "Louis was a true giant of a man," she said, "in brilliance, vision, and stature. And he had a great sense of humor. It took a year for Leakey to get the money. The British administration initially refused to grant permission, horrified at the thought of a young white woman going off into the bush, but Leakey persisted and in the end they agreed, provided I did not

Mum helped with pressing plants I collected that the chimpanzees were eating, as well as drying skulls and other bones that I found. We are in the entrance of our secondhand, ex-army tent. (JANE GOODALL INSTITUTE/HUGO VAN LAWICK)

go alone and had a 'European' companion. Louis wanted someone who would support me in the background, not compete with me, and decided Mum would be perfect. I don't think he had to twist her arm very hard. She loved a challenge. The whole expedition would not have been possible without her.

"Bernard Verdcourt, the botanist at the Coryndon Museum, drove us overland to Kigoma—the closest town to Gombe—in an overloaded, short-wheelbase Land Rover on mostly dirt roads that were rutted and potholed. He later admitted that when he dropped us off, he never expected to see either of us alive again."

Jane, however, was more concerned about how she could accomplish her mission than the potential dangers. Jane paused, and I prompted her to continue. "When you were at Gombe, you wrote a letter to your family saying, 'My future is so ridiculous, I just squat here, chimp-like, on my rocks, pulling out prickles and thorns, and laugh to think of this unknown *Miss Goodall* who is said to be doing scientific research somewhere.' Take me back to those moments of hope and hopelessness," I said, eager to understand the uncertainty and self-doubt that she faced, especially when trying to do something that had never been done before.

"There were so many moments of disappointment and despair," Jane explained. "Awakening before dawn each day, I would climb the steep hills of Gombe in search of chimpanzees, catching rare glimpses of them through my binoculars. I would creep and crawl through the forest, exhausted, my arms and legs and face scratched by the undergrowth, and finally I would come upon a group of chimpanzees. My heart would leap but before I could observe anything, they would take one look at me and run away.

"There was only six months of money to support my research,

and chimps were just running away from me. The weeks became months. I knew, given time, I could get the chimps' trust. But did I have the time? I knew if this did not happen, I would be letting Leakey down; he had put so much confidence in me, and the dream would come to an end. Yet most of all," Jane continued, "I would never be able to understand these fascinating creatures—or what they could tell us about human evolution, which is what Leakey was hoping to better understand."

Jane wasn't an established scientist. She did not even have an undergraduate degree. Leakey wanted someone whose thinking was not already compromised by too much academic prejudice or preconceived beliefs. Jane's breakthrough discoveries, especially about animal emotions and personalities, might never have been possible if she had been trained to deny that animals could have these, as was common in universities at the time.

It was fortunate for Jane that Leakey believed that women might make better field researchers—that they might be more patient and show more empathy toward the animals they were studying. After sending Jane into the forest, Leakey helped two other young women follow their dreams, finding funding for Dian Fossey to study mountain gorillas and Biruté Galdikas to study orangutans. The three women later became known as "the Trimates."

"When I saw the rugged, mountainous terrain at the park," Jane said, "I wondered how on Earth I would ever find the elusive chimpanzees, and it was not easy. Mum played a very important role. I would come back to camp depressed, because the chimpanzees had, again, just run away from me. But Mum would point out that I was learning more than I realized. I had discovered a peak where I could sit and overlook two valleys. And through my binoculars, I

I fixed a camera in a tree and took a timed photo of me searching for signs of chimpanzees. (JANE GOODALL INSTITUTE/JANE GOODALL)

watched as they made sleeping nests up in the trees and traveled in different-sized groups. I learned what foods they were eating and their different calls."

But Jane knew that it was not enough information to enable Leakey to get more money when the six-month grant ended.

"I wrote many letters to Leakey," Jane recalled, "when the chimps were running away: 'You've put all your faith in me and I can't do it.' And he'd write back and say, 'I know you can.'"

"Leakey's encouragement must have meant a lot to you."

"Actually, he made it worse," Jane insisted. "Every time he said, 'I know you can do it,' I was thinking, 'But if I don't, I've let him down.' That was what really concerned me. He'd stuck his neck out getting money for this young, unknown girl. And how would he feel, and how would *I* feel, if I let him down?" She wrote to him again and again, in desperation. "'It's not working, Louis,' I'd write. And he'd write back, 'I *know* you can do it.' In his next letter the word 'KNOW' was bigger and underlined. So I felt increasingly desperate."

"There must've been something in his belief that you could do it that also encouraged you to get back out there," I suggested.

"It probably encouraged me to work even harder, although I don't know how I could have worked harder, as I was going out every morning at half past five and crawling about the forest or watching from my peak all day until it was almost dark."

Those early days sounded like they were full of dangers, challenges, and obstacles. But Jane seemed undaunted. She told me how she once sat on the ground and watched a poisonous snake slither over her legs. And how she felt no animals would hurt her, as she was "meant to be there." She believed the animals would somehow know that she meant them no harm. Leakey had encouraged this belief and, thus far, no wild animal had ever harmed her.

As important as her belief, Jane also knew how to behave around wild animals. In particular, she knew the most dangerous thing was to get between a mother and her child, or to confront a wounded animal or one that had learned to hate humans. "Leakey approved of how I had reacted at Olduvai when one evening, after a hard day's work under the hot sun, Gillian and I were walking back to camp and I sensed something behind me—and there was a curious young male lion," Jane said. He was adult-sized, but his mane was only just beginning to grow. She told Gillian they should simply walk away slowly and climb up the side of the gorge onto the open plain above.

"Louis said it was lucky we hadn't run, or the lion might have chased us. He also approved of how I had reacted when we came upon a male black rhino. I said we should stand absolutely still, as rhinos don't see well; and luckily I could feel the wind blowing toward us, so I knew our smell would be carried away from him. The rhino knew there was something odd and ran back and forth with his tail in the air, but finally, he trotted away. I think that these reactions—and my willingness to dig for fossils eight hours a day—are probably why Leakey offered me the chance of studying the chimps."

In Gombe, Jane persevered, and slowly she won the trust of the chimpanzees. As she got to know them, she gave them names—as she had named every animal she had ever owned or watched. Later she would be told that it was more "scientific" to identify them by number. But Jane, never having been to college, did not know this—and she told me that even if she had, she is sure she would have named the chimpanzees anyway.

"David Greybeard, a very handsome chimpanzee with distinguished white hair on his chin, was the first to trust me," Jane said. "He was very calm and I think it was his acceptance of me that gradually persuaded the others that I was not so dangerous after all."

It was David Greybeard who Jane first observed using grass stems as tools to fish out termites from a termite mound—their earthen nest. And then she saw him stripping leaves from a leafy twig to make it suitable to use for the purpose. At the time Western science believed only humans were capable of making tools and that this was a main reason why we were separate from all other animals. We were defined as "Man the toolmaker."

When Jane's observations were reported, this challenge to human uniqueness caused a worldwide sensation. Leakey's famous telegram to Jane said: "Ah! We must now redefine man, redefine tools, or accept chimpanzees as human!" David Greybeard was eventually named one of the fifteen most influential animals that ever lived by *Time* magazine.

David Greybeard on a termite mound with a grass tool in his mouth, taken just after first sighting of termite fishing. (JANE GOOD-ALL INSTITUTE/JUDY GOODALL)

"David Greybeard and his tool using was the moment that changed everything," Jane recalled. "*National Geographic* agreed to take on funding my research when the first grant ran out, and they sent Hugo to film it all." Hugo van Lawick, the Dutch filmmaker who recorded Jane's discoveries, ultimately became her first husband.

"It was all thanks to Louis suggesting that Hugo would be the ideal person and the *Geographic* agreeing to send him," Jane said, referring to the ensuing romance.

"So Louis was the matchmaker?"

"Yes, he was. I wasn't actually looking for a 'mate,' but Hugo arrived in the middle of nowhere, and there we were. We were both reasonably attractive. We both loved animals. We both loved nature. So it was pretty obvious that it should have worked."

Shows the heavy equipment Hugo lugged around, an old Bolex 16 mm camera. On Gombe beach. (ABC NEWS PUBLICITY PHOTO)

Jane recalled her first marriage with the equanimity of almost five decades since their eventual divorce in 1974. She would marry again, to Derek Bryceson, the Tanzanian parks director, but would lose him to cancer less than five years later, when she was just forty-six.

When Jane went into the forest with her own hopes and dreams, she had no idea that hope itself would ultimately become a central theme of her work.

"What was the role of hope during those early days?"

"If I hadn't had hope that I could, given time, gain the chimps' trust I would have given up."

Jane paused and looked down. "Of course there was the nagging worry—*did* I have time? I suppose it's a bit like climate change. We know we *can* slow it down—we're just concerned as to whether we have sufficient time to effectively turn things around."

We both sat in silence, feeling the weight of Jane's question. Even before the climate crisis was well known, it was her concern for the chimpanzees and the environment that led her to leave Gombe.

"During my early days in Gombe I was in my own magical world, continually learning new things about the chimpanzees and the forest. But in 1986 everything changed. By then there were several other field sites across Africa and I helped to organize a conference to bring these scientists together."

It was at this conference that Jane learned that in every place where chimps were being studied across their range their numbers were dropping and their forests were being destroyed. They were being hunted for bushmeat, caught in snares, and exposed to human diseases. Mothers were shot so their infants could be taken and sold as pets or to zoos, or trained for the circus, or used for medical research.

Jane told me how she secured funding to visit six different countries across the chimpanzee range in Africa. "I learned a great deal about the problems facing the chimps," she said, "but I also learned about the problems facing human populations living in and around chimpanzee forests. The crippling poverty, lack of good education and health facilities, degradation of the land as their populations grew.

"When I went to Gombe in 1960," Jane said, "it was part of the great equatorial forest belt that stretched across Africa. By 1990, it had become a tiny oasis of forest surrounded by completely bare hills. More people were living there than the land could support, too poor to buy food from elsewhere, struggling to survive. Trees had been cleared to grow food or make charcoal.

"I realized that if we couldn't help people find a way of making a living without destroying the environment, there was no way we could try to save the chimpanzees."

I knew that Jane had spent the last three decades fighting. Fighting for the rights of animals, people, and the environment, and I was sobered when she added, "Now the damage we have inflicted is undeniable."

I finally gathered the courage to ask Jane the more personal question I had been hesitant to ask. "Have you ever lost hope?" I did not know if the world's icon of hope would admit to having ever lost it.

She paused and reflected on the question. I knew her drive and

When we were in Dar es Salaam, Derek and I contacted Gombe every day on the radio telephone, seen on the table. The rescue dog is Wagga. (JANE GOODALL INSTITUTE/COURTESY OF THE GOODALL FAMILY)

her resilience made it unlikely, but I also knew that she was no stranger to crisis and heartbreak. Finally, she exhaled. "Maybe, for a time. When Derek died. Grief can make one feel hopeless."

I waited for Jane to go on as we explored difficult memories.

"I will never forget his last words. He said, 'I didn't know such pain was possible.' I keep trying to forget what he said, but I can't. Even though there were times when he wasn't in pain, and he was okay, that doesn't drown out those last anguished words. It's horrible."

I imagined the heartache of hearing your spouse in such excruciating pain. "How did you manage?"

"After his death, lots of people helped me. I returned to the sanctuary of my home in England, the Birches," Jane said. "One of the dogs helped me a lot, too. She slept on my bed, providing the kind of comfort that I have always derived from the companionship of a loving dog. And then I went back to Africa and went to Gombe. It was the forest that helped most of all."

"What did you get from the forest?"

"It gave me a sense of peace and timelessness, and reminded me of the cycle of life and death we all go through . . . and I kept busy. That helps."

"I can only imagine how difficult that time must have been," I said. I had not yet lost anyone as close as a spouse or parent, but I was moved by the heartbreak in her words that still echoed across decades.

Bugs yawned and jumped down from Jane's lap, his nap done, ready for his next meal or his next adventure.

"Have you ever lost hope in the future of humanity?" I asked, knowing that hopelessness can be both deeply personal and also sweepingly global, especially as so many things seemed to be moving in the wrong direction.

"Sometimes, I think, 'Well, why on Earth do I feel hopeful?' Because the problems facing the planet are huge. And if I analyze them carefully, they do sometimes seem absolutely impossible to solve. So why do I feel hopeful? Partly, because I'm obstinate. I just won't give in. But it's partly also because we cannot accurately predict what the future might bring. We simply can't. No one can know how it will all turn out."

Somehow, hearing how Jane's hope had been tested and questioned made it more inspiring and, even in a strange way, more trustworthy.

Yet I wondered why some people bounce back faster than others from grief or heartbreak. Was there any science that could explain hope and why some people have more of it than others, and perhaps how all of us can have it when we need it?

Can Science Explain Hope?

When Jane and I agreed to work on a book about hope, I did some research into the relatively new field of hope studies. I was surprised to learn that hope is quite different from wishing or fantasizing. Hope leads to future success in a way that wishful thinking does not. While both involve thinking about the future with rich imagery, only hope sparks us to take action directed toward the hoped-for goal—something I would hear Jane emphasize repeatedly during our subsequent meetings.

When we focus on the future, we do one of three things. We *fantasize*, which involves big dreams that are mostly for fun and entertainment; we *dwell*, which involves focusing on all the bad stuff that might happen—this was the official pastime of my hometown—or we *hope*, which involves envisioning the future while recognizing the inevitability of challenges. Interestingly, more hopeful people actually anticipate setbacks along the way and work to remove them. I was learning that hope wasn't just a Pollyanna avoidance of the problems but a way of engaging with them. And yet I always imagined that hopeful and optimistic people are just born that way and wanted to know if Jane agreed.

"Aren't some people just more hopeful or more optimistic than others?"

"Well, maybe," Jane said, "but hope and optimism are not the same thing."

"What's the difference?"

"I haven't got the faintest idea," she said with a laugh.

I waited, knowing Jane loved scientific inquiry and debate. I could see she was considering the difference.

"Well, I guess a person either is or isn't an optimist. It's a disposition or a philosophy of life. As an optimist, you can just have the feeling, 'Oh, it'll be all right.' It's the opposite of a pessimist, who says, 'Oh, that's never going to work.' Hope, on the other hand, is a stubborn determination to do all you can to *make* it work. And hope is something we can cultivate. It can change over the course of our lifetime. Of course, someone with an optimistic nature is far more likely to be hopeful because they see a glass as half full rather than half empty!"

"Do our genes," I asked, "determine whether we are an optimist or a pessimist?"

"From all I've read," Jane said, "there is evidence that an optimistic personality may be partly the result of genetic inheritance, but this can surely be overruled by environmental factors—just as those born without a genetic tendency toward optimism can develop a more optimistic and self-reliant outlook. It certainly points to the importance of a child's environment and early education. A supportive family background can have a *major* effect—I know I was really lucky with mine, especially my mother. But how do we know that I would have been less optimistic with a less supportive family? I remember reading somewhere that a pair of identical twins, brought up in different backgrounds, still showed similar personalities. But as I said, it's also true that the environment can affect the expression of genes."

"Have you heard the joke about the difference between an optimist and a pessimist?" I asked. "The optimist thinks that this is the best of all possible worlds, and the pessimist fears the optimist is right."

Jane laughed. "We don't really know how it will all turn out, do we? And we can't just think that we can do nothing and everything will work out for the best."

Jane's pragmatic view made me think of a conversation I had had with Desmond Tutu, who had endured so many tragic reversals and so much adversity in the struggle to free South Africa from the racist apartheid regime.

I recalled to Jane, "Archbishop Tutu once told me that optimism can quickly turn to pessimism when the circumstances change. Hope, he explained, is a much deeper source of strength, practically unshakable. When a journalist once asked Tutu why he was optimistic, he said he was not optimistic, he was a 'prisoner of hope,' quoting the biblical prophet Zechariah. He said hope is being able to see that there is light despite all of the darkness."

"Yes," Jane said. "Hope does not deny all the difficulty and all the danger that exists, but it is not stopped by them. There is a lot of darkness, but our actions create the light."

"So it seems we can shift our perspective to see the light and also to work to create more of it."

Jane nodded. "It is important to take action and realize that we *can* make a difference, and this will encourage others to take action, and then we realize we are not alone and our cumulative actions truly make an even greater difference. That is how we spread the light. And this, of course, makes us all ever more hopeful."

"I'm always a little skeptical," I said, "of attempts to quantify something as intangible as hope, but there seems to be some interesting research that hope has a profound impact on our success, happiness, and even health. One meta-analysis of over a hundred hope studies found that hope leads to a twelve percent increase in academic performance, a fourteen percent increase in workplace outcomes, and a fourteen percent boost in happiness. What do you think of all that?"

"I am sure hope makes a significant difference in so many aspects of our life. It impacts our behavior and what we are able to achieve," Jane said. "But I think it's also important to remember that while statistics can be helpful, people are moved to action by stories more than statistics. So many people thank me for not putting statistics in my lectures!"

"But don't we want to tell people the facts?" I asked.

"Well, let's put them in the back of the book for those who want all the detail."

"Okay, we can add a section of Further Reading for those who want to learn more about the research that we discuss in the dialogue," I said, and then asked Jane about the communal nature of hope. "What do you think is the relationship between the hopeful-ness that people feel in their own lives and their hopefulness about the world?"

"Let's say you're a mother," Jane replied. "You hope that your child is going to be well educated, get a good job, be a decent per-son. You hope that in your life you're going to be able to get a good job and support a family. That's for you and your life. But your hopes are obviously extended to hope for the community and country you live in. Hoping your community will fight a new de-velopment that will pollute the air and affect your child's health. Hope that the right political leaders will be elected to make your hopes easier to attain."

It was clear, as Jane was saying, that each of us has our hopes and dreams for our own life and our hopes and dreams for the world. Hope science has identified four components that are essential for any lasting sense of hope in our lives—and perhaps in our world. We need to have realistic *goals* to pursue as well as realistic *pathways* to achieve them. In addition, we need the *confidence* that we can achieve these goals, and the *support* to help us overcome adversity along the way. Some researchers call these four components the "hope cycle" because the more of each we have, the more they encourage each other and inspire hope in our life.

The science of hope was interesting, but I wanted to know what Jane thought, especially about how we could have hope in trou-bled times. But before we could explore this question, Dr. Anthony Collins, Jane's colleague in Gombe, came in and told us that the *National Geographic* film crew needed Jane. We stopped for the day and agreed to pick up the next morning to discuss hope in the face of crises. Little did I know that by the next night, hope would suddenly become even more urgent—and elusive—as I faced a crisis of my own.

How Do We Have Hope in Trying Times?

I was awoken early in the sticky heat of the Tanzanian summer morning by the murmuring of the muezzin call to prayer. In the pink dawn light, as the blue water and blue sky were brightening, I looked out at a fisherman in a tiny wooden boat, more like a dugout canoe, who was throwing a delicate white net on the water, hoping to catch a fish. He threw it again and again, each time pulling it in and plucking out the sticks and leaves and occasionally a plastic bag and bottle that he caught, but no fish. Surely, it was hope—and hunger—that got him up each morning to feed his family.

When I arrived at Jane's house later that morning, she met me in the back garden and pointed to a dark stain on the knee of her pants.

"It's blood," she said. As we walked in her large and wild garden, Jane showed me where she had tripped the night before and cut her knee.

She explained how it happened. "I was holding the candles up here," she said, raising her hands high, "so I could see where I was headed, but not the ground below. Someone said, 'Mind the step,' but by that time I was already flat out."

Jane seemed unruffled by her injury.

"My body heals fast," she said.

"You've had worse, I'm sure," I said, trying to reflect her keep-calm-and-carry-on attitude.

"Oh yes. Look at this," Jane said, pointing to her cheek, almost relishing the dent that was a likely sign of a cracked bone.

"What was that?"

"It was an interaction with a rock at Gombe."

"Tell me what happened."

"Well, if we're going to talk about it, I'll tell you in detail, because it was dramatic—"

But before she could begin, the dogs ran up and jumped up on us affectionately. One, Marley, was a small white dog with short legs, something like a cross between a corgi and a West Highland terrier, with high tufted ears. The other, Mica, was a larger brown and black mix with the floppy ears of a Lab.

"All rescued," Jane said. "Mica came from a shelter started by a friend. And Merlin found Marley wandering on the streets, homeless. We have no idea of their past history." She petted them as she began her story.

"It happened twelve years ago, when I was seventy-four. I was climbing up a slope that was really too steep. It was silly of me, but the chimp had gone somewhere up there and I wanted to try to find her. It was skiddy and the dry season, and there was nothing much at the side to grab on to, just bits of dry grass that came away in my hand. However, I got nearly to the top, and there just above me, was this big rock, and I thought I'd just pull myself onto that rock and then onto another that I could see above it—and then I'd be up. So I reached up, grabbed hold of the rock, and to my horror, it just came out of the earth. And it was about that big"—Jane held her hands two feet apart—"and it was very, very solid and heavy. And so it landed on my chest and we tumbled down together—I think I ended up on my side, sort of clutching the rock to me! The slope, like I said, was steep and about thirty meters—a hundred feet—to the bottom. If something hadn't pushed me to the side, into some vegetation that I didn't even think was there, I wouldn't be here now. I was saved but the rock went all the way to the bottom. It took two men with a stretcher to carry it back. It was too heavy for me to lift. We have it outside my house in Gombe," Jane concluded, describing her trophy triumphantly. "We make people guess how heavy it is."

"How much does it weigh?" I asked.

"One hundred pounds or fifty-nine kilos."

"But if you add the velocity of falling, that must have had a much greater impact on your body as you were tumbling down the slopes," I said.

"Don't I know it!" Jane said.

"What was it that pushed you to the side?"

"Somebody or some unknown power looking after me up there," Jane said, glancing up. "That sort of thing has happened before."

"Somebody—" I began, but Jane was still in the midst of her story. We did not have a chance to discuss who or what was looking after her, but I was sure we would return to the subject later.

"So when I got to an X-ray machine two days later I found I had a dislocated shoulder. And much later after the bruising had long gone from my face, I was sure there was something wrong. And I asked my dentist if he could do an X-ray."

"Your *dentist*?"

"Yes, well, you see I was there anyway—and I didn't want to go through all the bother of making a doctor's appointment. He said he couldn't do a proper X-ray, but it looked as though I had a cracked cheekbone. 'They could put in a metal plate,' he said. But I was quite sure I didn't need a plate in my cheek. Just think of security at the airport! And, anyway, I didn't have time for aches and pains. I had a job to do. I still don't have time for aches and pains. I still have a job to do."

So many older people I knew spent a great deal of time focused on their aches and pains, but those who seemed healthiest and happiest were those who focused on something beyond their own troubles. Jane was revealing a powerful example of the resilience and persistence in the face of obstacles that the researchers had told me was essential for hope. Nothing was going to get in the way of her reaching her goal.

"Were you always so strong and so tough?" I asked.

Jane laughed. "No, I was always getting sick when I was young. In fact my uncle Eric, a doctor, used to call me Weary Willy. And I honestly used to think that my brain rattled around in my head. I can't think why. But I really did get terrible migraines."

"I used to get migraines, too. They're awful," I said.

I was impressed by her mental fortitude that seemingly willed her to become so hardy in adult life. It reminded me of one of the most moving stories I had heard about the power of the mind.

"Do you know the work of psychologist Edith Eger?" I asked, knowing Jane's fascination with the Holocaust and what it reveals about human nature.

"No, tell me who she is."

"Dr. Eger was just sixteen when she was in a cattle car with her family on the way to Auschwitz. Her mother said to her, 'We don't know where we're going. We don't know what's going to happen. Just remember, no one can take away from you what you've put in your mind.' She remembered her mother's words even after her parents were sent to the crematoria.

"While everyone around her, from the guards to the other inmates, told her she would never get out alive, she never lost hope. She told herself, 'This is temporary. If I survive today, tomorrow I will be free.' One of the other girls in the death camp was very ill. Every morning Dr. Eger expected to see the girl dead in her bunk. Yet every day she managed to raise herself off her wooden bed and to work another day. Each time she stood in the selection line, she managed to look healthy enough that she was not sent to the gas chamber. Each night she collapsed back onto her bunk, gasping for breath.

"Edie asked her how she managed to keep going. The girl said, 'I heard we're going to be liberated by Christmas.' The girl counted down each day and each hour, but Christmas came and they were not liberated. She died the next day. Edie says that hope kept the girl alive and that when she lost hope, she lost her will to live.

"She says that people who wonder how you can have hope in seemingly hopeless situations, like a death camp, confuse hope with idealism. Idealism expects everything to be fair or easy or good. She says it's a defense mechanism not unlike denial or delusion. Hope, she says, does not deny the evil but is a response to it." I was beginning to see that hope was not just wishful thinking. It did take the facts and the obstacles into account, but it did not let them overwhelm or stop us. Certainly, this was true in many seemingly hopeless situations.

"I know," said Jane thoughtfully, "hope isn't always based on logic. In fact, it can seem very illogical."

The global situation today certainly could seem hopeless, and yet Jane was feeling hope even when "logic" might tell us that there was no reason for it. Maybe hope is not an expression of the facts alone. Hope is how we create new facts.

I knew that Jane's hopefulness in spite of grim global realities was focused around four main reasons for hope: the *amazing human intellect*, the *resilience of nature*, the *power of youth*, and the *indomitable human spirit*. And I knew that she traveled the world sharing this wisdom and inspiring hope in others. I was eager to explore and debate them with her. Why did she think that our amazing human intellect was a source of hope, given all the evil it was capable of doing? Had our cleverness not brought us to the brink of destruction? I could imagine why she found hope in the resilience of nature, but could it possibly withstand the destruction we were unleashing? And why were young people a source of hope for her, given that previous generations had not been able to solve the problems we face and that the youth were not yet ruling the world? And finally, what did she mean by the indomitable human spirit, and how could it possibly save us? But our time together for the day had ended, and we agreed that we would pick up our discussion early the next morning.

But our plans were about to be disrupted.

Late that night, my cell phone rang. It was my wife, Rachel. My father had been rushed to the hospital, and the situation looked serious. I booked the next flight to New York and called Jane to tell her that I would need to postpone our talks until my father was stable. For me, hope and hopelessness were no longer intellectual. They were everything.

(CATALIN AND DANIELA MITRACHE)

II

Jane's Four Reasons for Hope

REASON 1:
THE AMAZING HUMAN INTELLECT

Freud was the alpha male at this time. Intelligent and an excellent boss. Will we ever know what they are thinking? (MICHAEL NEUGEBAUER/WWW .MINEPHOTO.COM)

"I was so sorry to hear about your father," Jane said four months later when we met in the Netherlands.

What had originally been diagnosed as routine aging and

increasing infirmity in my father's legs turned out to be aggressive T-cell lymphoma of the central nervous system attacking his spinal cord and then his brain. I spent the months after I flew back from Dar es Salaam visiting him at the hospital in New York as he heroically attempted to hold on to hope and his consciousness until both ultimately succumbed to the cancer. I will never forget the bravery and equanimity with which he met the news that his cancer was incurable and that he only had weeks or possibly months to live. "I guess it's time to face the inevitable," he said.

When he was in excruciating pain and on the precipice of death, I asked him how long he thought he would stay around. "Just as long as it takes to get landing instructions," he said, "or to leap into eternity." I was seeing that there were limits to hope, and I was still deep in grief. Jane's kindness and understanding meant a great deal during the brutal months of my dad's illness and death.

Jane and I were meeting now in a renovated forester's cabin in the middle of the woods in a nature preserve near Utrecht. The cottage was cozy and well insulated against the bone-chilling cold and wind that blows across much of the Netherlands in the winter. We sat across from each other as slanted sunlight filtered through the windows and a log fire crackled.

Jane had just spent four days at her family home in England after a long trip to Beijing, Chengdu, Kuala Lumpur, Penang, and Singapore. Despite her almost constant travels around the world, she seemed energetic and eager to begin. She wore a blue turtleneck and a green jacket, and her hands were clasped across a gray wool blanket.

"Thank you for your condolences," I said. Jane had written when my father finally passed. "I'm sorry I had to leave so suddenly."

Doug's father, Richard Abrams, several years
before his diagnosis. (MICHAEL GARBER)

"The sad part was the reason you had to go."

"It's been a difficult few months," I admitted.

"You don't really get over it. It is such a great loss," Jane said. "I
guess the depth of our grief is a reminder of the depth of our love."

I smiled, touched by her words. "He was a wonderful father."

On his deathbed it was his heart and his love even more than his
head and his reason that seemed most important to him, so I wondered why the human intellect was one of Jane's reasons for hope. If
anything, as my father's neurons fired together and then unraveled
into delirium, I'd been struck by how fragile our consciousness is.
Our mind seemed so delicate and so fallible.

We sat for a moment, honoring the memory of my father and all
the people we had lost. And then we began.

"Why is the human intellect one of your reasons for hope?" I asked.

From Prehistoric Ape to Master of the World

"Well, it is what makes us most different from chimpanzees and other animals," Jane said, "the explosive development of our intellect."

"What exactly do you mean by the human *intellect*?"

"The part of our brain that reasons and solves problems."

At one time scientists said that these characteristics were confined to humans, and Jane was one of the people most responsible for showing that intelligence is on a continuum across all animals, including humans. I mentioned this to Jane.

"Yes, today we know that animals are way, way more intelligent than people used to think," she said. "Chimpanzees and the other great apes can learn four hundred or more words of American Sign Language; work out complex problems on a computer; and, like some other animals including pigs, love painting and drawing. Crows are amazingly intelligent, as are parrots. Rats are incredibly smart."

"I remember you also telling me in Tanzania that octopuses are extraordinarily clever and can solve all sorts of problems even though their brains are structured so differently from mammalian brains," I said.

Jane laughed. "They actually have brains in each of their eight arms! Here's something else you'll like—apparently you can teach bumblebees to roll a little ball into a hole for the reward of a drop of nectar. And even more remarkably—other bees, who have not been trained, can perform the same task after merely watching the trained bees. We are learning new things all the time and

Many animals express their intelligence through art. Pigcasso, originally destined to become ham and bacon, was rescued by Joanne Lefson and taught to paint. She likes to paint with a good view—in the background you see Table Mountain. Her paintings sell for thousands of dollars. (WWW.PIGCASSO.ORG)

I always tell students that this is a wonderful time to study animal intelligence."

"So what exactly about our intellect makes us different from all other animals?" I asked.

"Even though chimpanzees, our closest living relatives, can perform super well at all kinds of intelligence tests, even the brightest chimpanzee could not design that rocket from which crept a robot that was programmed to crawl around the surface of the Red Planet— Mars—taking photos for scientists on Earth to study. Humans have done such incredible things—I mean, think of Galileo and Leonardo da Vinci and Linnaeus and Darwin and Newton and his apple; think of the pyramids and some of the great architecture and our art and music."

Jane paused, and I started thinking of all the brilliant people who had come up with theories and constructed fabulous buildings with none of the sophisticated tools we have today and who had access to none of the knowledge that has been built up from the past. Jane broke into my musings.

"And you know, Doug, whenever I look at a full moon up in the sky, I get the same feeling of awe and wonder that I felt on that historic day in 1969 when Neil Armstrong became the first human to step on the moon and was closely followed by Buzz Aldrin. And I think to myself, 'Human beings have actually walked up there. Wow!' When I give talks I always tell people, when they next look at the moon, try to capture that feeling of awe—don't just take it for granted.

"So, yes," Jane continued, "I honestly think it was the explosion of the human intellect that took a rather weak and unexceptional species of prehistoric apes and turned them into the self-appointed masters of the world."

"But if we're so much more intelligent than other animals, how come we do so many stupid things?" I asked.

"Ah!" said Jane. "That's why I say 'intellectual' rather than 'intelligent.' An *intelligent* animal would not destroy its only home—which is what we have been doing for a very long time. Of course, some people are indeed highly intelligent, but so many are not. We labeled ourselves Homo sapiens, the 'wise man,' but unfortunately there is not enough wisdom in the world today."

"But we are clever and creative?" I said.

"Yes, we are very clever and very creative. And like all primates and many other animals, we are very curious creatures. And our curiosity, coupled with our intellect, has led to many great discoveries in many fields because we like to find out how and why things work the way they do, pushing the boundaries of our understanding."

"So what do you think made the difference?" I asked. "Why did human brains evolve beyond chimp—?"

"Language," Jane said, anticipating my question. "At some point in our evolution, we developed this ability to communicate with words. Our mastery of language allowed us to teach about things that weren't present. We could pass on wisdom gleaned from the successes and mistakes of the past. And we could plan for the distant future. Most importantly, we could bring people together to discuss problems, people from different backgrounds with different knowledge."

I was intrigued to hear that Jane believed that language had led to the explosion in the human intellect. Because interestingly, while researching hope, I discovered that language, goal setting, *and hope* all seem to arise in the same area of the brain—the prefrontal cortex, which is right behind our forehead and is the most recently evolved part of our brain. This region is larger in humans than in other great apes.

For a while we talked about all that humans have accomplished, such as designing machines that enable us to fly through the air and travel under the ocean, and technology that lets us communicate with people on the other side of the globe almost instantaneously.

"So it is really strange, isn't it, that this amazing human intellect has also gotten us into this mess we're in," I said. "This very same intellect has created a world out of balance. One could argue that the human intellect was the greatest mistake in evolution—a mistake that is now threatening all life on the planet."

"Yes, we have certainly made a mess of things," Jane agreed. "But it's the *way we have used the intellect* that has made the mess, not the intellect per se. It's a mixture of greed, hate, fear, and desire for power that has caused us to use our intellect in unfortunate ways. But the good news is that an intellect smart enough to create nuclear weapons and AI is also, surely, capable of coming up with ways to heal the harm we have inflicted on this poor old planet. And indeed, now that we've become more and more aware of what we've done, we have begun to use our creativity and inventiveness to start healing the harm we've caused. Already there are innovative solutions, including renewable energy, regenerative farming and permaculture, moving toward a plant-based diet, and many others, that are directed toward creating a new way of doing things. And, as individuals, we are recognizing that we need to leave a lighter ecological footprint and we are thinking of ways to do it."

"So our intellect in itself is neither good nor bad—it depends on how we humans choose to use it—to make the world better or to destroy it?"

"Yes, that's where our intellect and our use of words makes us different from other animals. We are both worse and better, because we have the ability to choose." Jane smiled. "We're half sinner and half saint."

Half Sinner, Half Saint

"In the end, which wins out—the evil or the good?" I asked. "Are we fifty-one percent good or fifty-one percent evil?"

"Well, there's plenty of evidence for both sides of this debate, but I think we're split down the middle," Jane said. "Humans are incredibly adaptive and will do whatever is required to survive in their environment. The environment we create will determine what prevails. In other words, what we nurture and encourage wins."

There's a strange feeling of having your world turned upside down. I was experiencing that disorienting feeling of seeing the world in a new way.

What I had been calling good and evil were simply the qualities of kindness and cruelty, generosity and selfishness, tenderness and aggression that we had evolved to survive in different environments and under different circumstances. And as Jane had said, we will do whatever it takes to survive in the world. If we live in a society with a reasonable standard of living and some degree of social justice, the generous and peaceful aspects of our nature are likely to prevail; while in a society of racial discrimination and economic injustice, violence will thrive.

"Well," said Jane, when I shared these thoughts, "I think that to a large extent that's true. Think of the genocide that occurred both in Rwanda and Burundi where the ethnic mix of Hutu and Tutsi is the same. International aid poured into Rwanda after the genocide because of President Bill Clinton's visit. But Burundi was more or less ignored. As a result, Rwanda has been able to build up its infrastructure with roads and hospitals, international businesses have established themselves, and the Hutu and Tutsi are living seemingly in peace. Whilst in Burundi, none of this has happened and there is still periodic violence and bloodshed to this day.

"But we have to remember that a society is made up of people, and there are always people seeking change. So many Burundian citizens are wanting to create a more peaceful society. Societies only *seem* stable when they are ruled over by an autocratic government.

Think of the ethnic conflicts that emerged after the breakup of the Soviet Union."

"Do you think we're capable of a peaceful and harmonious society? What about our aggressive tendencies?"

Jane shook her head. "Aggressive behavior has almost certainly been part of our genetic makeup inherited from our far distant hominid ancestors. You know the reason Leakey sent me off to Gombe was because he believed humans and chimpanzees had a common ancestor about five to seven million years ago and that if I found behavior similar—or the same—in modern chimps and modern humans, it probably originated in that apelike humanlike ancestor and stayed with us during our separate evolutionary paths. And that would give him a better idea of how early humans—whose fossilized remains he had discovered in various parts of Africa—might have behaved. Things like kissing, embracing, and the bonding of family members. And—what is relevant to the question you asked—very similar aggressive patterns including a kind of primitive warfare between neighboring groups."

I remembered Jane telling me that she had been advised to play down the chimps' aggressive behavior because back then, in the 1970s, many scientists were trying to convince people that aggressive behavior was learned—the nature-versus-nurture controversy.

"Fortunately, because of our extraordinary intellect and our ability to communicate with words," Jane continued, "we have been able to progress beyond the purely emotional aggressive responses of other animals and, as I said, we have the ability to make conscious choices as to how we react in different situations. And the choices we make will partly reflect what we learned as children—and that will depend on the country and culture into which we were born.

"I suspect that it's true of small children everywhere that, when angry, they are likely to hit out at what has upset them. My sister, Judy, and I were taught that it was wrong to hit and kick and bite other children. In this way we acquired an understanding of the moral values of our society: this is good, and this is bad; this is right, and this is wrong. The bad and the wrong were punished—verbally—and the good and the right were rewarded."

"So children learn the moral values of their society," I said.

"Yes, and this makes human aggression worse than that of other species because we may act aggressively knowing full well that it is morally wrong—at least what we believe is morally wrong. For this reason I believe that only humans are capable of true evil—only we can sit down and, in cold blood, work out ways to torture people, to inflict pain. Carefully plan horrific cruelty."

I knew that this was a subject that haunted Jane. She had grown up in England while the Holocaust was happening in German-occupied Europe and had been shocked to discover its horrors. She was in Gombe when the genocides occurred in Rwanda and Burundi. Burundi is just north of Gombe. Some people, near the border between Tanzania and Burundi, said they saw the blood of slaughtered Burundians in the lake; and many refugees fleeing the violence in Burundi had settled in the hills behind Gombe. She had listened to horrifying stories of the barbaric cruelty from which they were fleeing.

Jane was also in Gombe when an armed group from the Democratic Republic of the Congo (DRC) abducted four of her students in the middle of the night. Much later, she had been in Kinshasa, the capital of the DRC, when there was rioting in the street outside the house where she was staying and a soldier was killed just below her window. She was in New York on 9/11 when terrorists flew planes into the Twin Towers.

She had stared deeply into the face of evil—she understood only too well the dark side of our nature. But Jane being Jane was always quick to see the wider perspective.

"Still," she said, as though addressing her own dark thoughts, "even though there is a lot of violence and evil in the world, from a historical perspective we can see a lot of change for the good. Just think—we're in the Netherlands. Less than a hundred years ago this land was soaked in blood during World War Two as the British fought the Germans. Recently, I was with some German friends, and I said, 'Isn't it weird that here we are—great friends, and our fathers were killing each other?' Now we have the European Union," Jane said. "All those countries that were warring with one another for hundreds of years are now united for the common good. It's a big sign of hope. Yes, we have had Brexit, which is a step backward, but we are unlikely to have war within the EU anytime soon."

I was inspired by Jane's hopefulness about the direction of human history and our increasing ability to prevent large-scale wars.

"But aren't you worried by the fact that authoritarian strongmen are rising all over the world right now?" I asked. "And by all the internal conflicts, the rise of nationalism. Even fascism is gaining traction—the neo-Nazis are getting stronger in America and, incredibly, in Germany. And on top of that there are so many conflicts going on in the world, so much violence—school shootings, gang warfare, domestic violence, and racism and sexism. How can you possibly be hopeful for the future?"

"Well, first of all, over the couple of million years that we have been humans, I do think we have increasingly become more caring and compassionate. And although there is much cruelty and injustice everywhere, there is general agreement that these behaviors are wrong. And more people understand what is going on thanks to the media. And when all's said and done, I do honestly believe that a far greater percentage of people are basically decent and kind.

"And there's another thing, Doug. Just as only we are capable of true evil," Jane said, "I think only we are capable of true altruism."

A New Universal Moral Code

"A chimpanzee," Jane continued, "will try to help another who is in trouble, but I think only we can perform an altruistic action even though we know it may harm us. Only we can decide to help someone in spite of knowing that it could put us in danger. It is truly altruistic when you help someone even when your intellect knows the risk you run. Think of the Germans who helped the Jews escape Nazi Germany, even hiding them in their houses. They knew it meant death if they were caught—and often they were."

"There was the theory of sociobiology that was very popular among scientists in the 1970s that explained altruism as merely a way of ensuring the survival of your own genes," I said. "So if you die helping family members, it's fine because your genes will survive into future generations. But I seem to remember you did not agree with that?"

"Well, though it's true so far as it goes," Jane said, "the research was based on helping behavior in social insects. Whereas we humans not only help our own kin but others in our group. And we also help individuals to whom we could not possibly be related.

"When it was realized that other animals also helped nonrelatives,

the next theory proposed was reciprocal altruism—you help someone in the hope that one day they will help you. But though these
theories may explain the evolutionary origin of altruistic behavior,
our intellect and imagination seem to allow us to be altruistic in a
more inclusive way. We humans help others even if it has no obvious
positive impact on our own lives. When we see a picture of starving
children, we can imagine how they feel and we want to help them.
The picture triggers our pity, our sympathy. And most people feel
this way even when those arousing this pity are from a different culture from their own. Photos—or even descriptions—of refugees from
war huddled in the winter in flimsy tents, or victims of earthquakes
who are starving and homeless, create a sort of visceral emotion. It
hurts—psychologically, that is. And it doesn't matter whether they are
European, African, or Asian, young or old. I remember sobbing when
I first read *Uncle Tom's Cabin*. How I hated the cruel slave owner and
all the others who inflicted this kind of pain and misery. Just as I hated
the German Nazis during the war."

After a pause Jane told me that it was only at that moment, as we
sat there in the hut in the Dutch wood, that she suddenly understood
how a sense of compassion for the victims of oppression can lead to
a hatred of the oppressor—and there you have a recipe for reciprocal
violence and internecine war like in Rwanda and Burundi.

"Are you saying we have to find a way of forgiving the oppressor?" I asked, somewhat suspect of this ability to forgive or have
compassion for the oppressor.

"Yes, I suppose we do. We have to think about the way they were
brought up, the code of ethics they were taught as children."

I referenced how Archbishop Tutu had chaired the Truth and
Reconciliation Commission in South Africa to avoid his country de-

volving into civil war. He had said that forgiveness is how we unchain ourselves from the past. We choose the forgiveness cycle instead of the revenge cycle.

"But you see, Doug," Jane continued, suddenly animated again, "this just shows you the importance of language. We can discuss these problems. We can teach our children the importance of looking at a problem from different points of view. To keep an open mind. To choose forgiveness instead of revenge."

The daylight was waning. I tried to read the expression on Jane's face, but it was shadowed in the dim light. I felt like Jane was leading me step-by-step to a deeper understanding of how we actually might find a better path ahead—though I was also skeptical of any easy solution.

"So what needs to happen?" I asked. "How do we evolve into better, more compassionate, more peaceful creatures?"

Jane poured the wee drams as she considered my question.

"We need a new *universal* moral code." Jane suddenly laughed. "I've just thought—every single major religion gives lip service to the Golden Rule—*Do unto others as you would have them do unto you.* So it's easy—there's our universal moral code. We just have to find a way to persuade people to honor it!" And then she sighed. "It does seem impossible, doesn't it, given all our human failings. Greed. Selfishness. Lust for power and wealth."

"Yes," I said—and then, tongue in cheek, said, "after all, we're only human!"

Jane took a sip.

She laughed, then added, "But I do honestly think we're moving in the right direction."

"So you really do think that we are becoming more caring?"

"I honestly think, Doug, that the majority of people are. Unfortunately, the media devotes so much space to covering all of the bad, hateful things that are going on and not enough to reporting about all the goodness and kindness that's out there. And think of it from a historical perspective. It wasn't long ago that women and children in England were forced to work in the mines in horrific conditions. Children went barefoot in the snow. Slavery was accepted and justified in many parts of the United States—in Britain, too.

"Admittedly, there are still many children living in poverty, and there is still slavery in many parts of the world, and there is racial and gender discrimination and unjust wages and so many other social ills—but more and more people are believing that all these things are not morally acceptable, and many groups are working hard to address all these issues and more. The apartheid regime has ended in South Africa. British colonial subjugation ended as the British Empire collapsed. Gradually the attitude toward women in many countries has changed. I was astonished the other day to see how many women have attained important positions in governments around the world. And there are so many lawyers standing up against injustice, speaking out for human rights—and in more and more countries lawyers and special-interest groups are fighting for animal rights as well."

I thought about this. Indeed, all that Jane had said represented steps forward toward a better global ethic. But I couldn't help thinking how many steps backward we have taken in recent years, and how much further we still had to go. I shared these thoughts with Jane, mentioning the horrific way immigrant children were being separated from their parents at the Mexican-US border, put into what amounts to cages, then sent off to "schools" in the deserts. And the rise in homelessness and the number of people who go to bed hungry. "And," I added, "we've already touched on the disturbing rise in nationalism."

"Yes, I know," Jane said. "And it is much the same in the UK and many other countries, too. It really is depressing."

I said, "I think this is what President Barack Obama meant with his famous statement that history 'zigs and zags' instead of moving forward in a straight line."

"It's easy to feel we are zagging backward," Jane said. "But it's important that we think about the protests that have succeeded and campaigns that have achieved their goals. Thanks to the internet . . ."

I was about to interrupt Jane, but she started to laugh. "Yes, I know all about the downside of this technology and especially about 'fake news'! But like our intellect, social media in itself is neither good nor bad—it is the use to which we put it that counts."

I had once asked Archbishop Tutu, whose stand against apartheid had bent the arc of history toward justice in South Africa, what he thought of human progress. It was just after the Paris bombings, and many people were despairing about humanity. He said that history takes two steps forward and one step back. Almost exactly a month later the world came together to ratify the Paris Climate Agreement. I'll never forget the other thing he said: "It takes time for us to become fully human." Perhaps he meant it takes time for us to evolve morally.

Jane thought about this for a bit. "I think maybe it takes a lot of time in our evolution for us to realize that we can never attain our full human potential unless our head and heart work together. It was that genius Linnaeus who gave our species the name of Homo sapiens, the *wise* human—"

"Clearly," I interrupted, "we are not living up to our name. You've already said that we are intellectually clever but not wise, so how do you understand 'wisdom'?"

The Wise? Ape

Jane was pensive for a moment, gathering her thoughts. "I think that wisdom involves using our powerful intellect to recognize the consequences of our actions and to think of the well-being of the whole. Unfortunately, Doug, we have lost the long-term perspective, and we are suffering from an absurd and very *unwise* belief that there can be unlimited economic development on a planet of finite natural resources, focusing on short-term results or profits at the expense of long-term interests. And if we carry on like this—well, I don't like to think what will happen. And it is most definitely *not* the behavior of a 'wise ape.'

"When making decisions most people ask, 'Will it help me or my family now or the next shareholders' meeting or my next election campaign?' The hallmark of wisdom is asking, 'What effects will the decision I make today have on future generations? On the health of the planet?'

"And it's the same sort of lack of wisdom shown by those in power who suppress certain sections of society—I mean, in America and the UK it is shameful the way certain sections of society are deliberately kept undereducated and underserved. And then the time comes when the resentment and anger of those people finally erupts, and they demand change. They want better wages or better health care or better schools. And this can lead to violence and bloodshed. Think of the French Revolution. And it was people fighting to end slavery that led to the American Civil War. Well, you and I both know of many stories of angry people coming together throughout history and using violence to overthrow oppressive political or social structures."

I thought about the costs of our lack of wisdom and our attempts to restore and heal from our mistakes. I asked Jane, "Do you think we will ever use our intellect in the right way?"

"Well, I don't think there will ever be a time when *everybody* will use it in the right way. There will always be sinners amongst us! But, as I keep saying, more and more people are fighting for justice and, by and large, I think humanity at least has a shared understanding of what justice means."

We flipped on more lights in the room, as it was now pitch-dark outside and the embers were getting low. Our whisky was finished, but we had one more riddle to solve. How could we use this amazing human intellect wisely? I put this question to Jane.

"Well, if we are ever going to do that—and I've already said that I think head and heart must work together—now is the time to prove that we can. Because if we don't act wisely *now* to slow down the heating of the planet and the loss of plant and animal life, it may be too late. We need to come together and solve these existential threats to life on Earth. And to do so, we must solve four great challenges—I know these four by heart because I often speak about them in my talks.

"First—we must alleviate poverty. If you are living in crippling poverty, you will cut down the last tree to grow food. Or fish the last fish because you're desperate to feed your family. In an urban area you will buy the cheapest food—you do not have the luxury of choosing a more ethically produced product.

"Second, we must reduce the unsustainable lifestyles of the

affluent. Let's face it, so many people have way more stuff than they need—or even want.

"Third, we must eliminate corruption, for without good governance and honest leadership, we cannot work together to solve our enormous social and environmental challenges.

"And finally, we must face up to the problems caused by growing populations of humans and their livestock. There are over seven billion of us today, and already, in many places, we have used up nature's finite natural resources faster than nature can replenish them. And by 2050 there will apparently be closer to ten billion of us. If we carry on with business as usual, that spells the end of life on Earth as we know it."

"Well, those are daunting challenges," I said.

"Yes, they are, but they are not insurmountable if we use our human intellect—together with good old common sense—to solve them. And, as I said earlier, we are beginning to make progress. Of course, a great deal of our onslaught on Mother Nature is not really lack of intelligence but a lack of compassion for future generations and the health of the planet: sheer selfish greed for short-term benefits to increase the wealth and power of individuals, corporations, and governments. The rest is due to thoughtlessness, lack of education, and poverty. In other words, there seems to be a disconnect between our clever brain and our compassionate heart. True wisdom requires both thinking with our head and understanding with our heart."

"Does some of our wisdom get lost when we lose connection with the natural world?" I asked.

"I believe it does. Indigenous cultures have always had a close connection with the natural world. There are so many wise shamans

We are learning wonderful things about trees, how they communicate under the ground and even help each other. (JANE GOODALL INSTITUTE/ CHASE PICKERING)

and healers among the indigenous people, so much knowledge about the benefits of living in harmony with the natural world."

"What have we forgotten—or chosen to ignore?"

"That there's intelligence in all life," Jane said. "I think indigenous people sense this when they talk about animals and trees being their brothers and sisters. I like to think that our human intellect is part of the Intelligence that led to the creation of the universe. Take the trees! We now know that they can communicate information to one another through underground networks of roots and the thin white threads of the microfungi that are attached to them."

I was familiar with the work of Suzanne Simard, one of the ecologists who had made this fantastic discovery. She called this network the Wood Wide Web because the trees of a forest are all connected under the ground. And that through this network, trees can receive information about their kinship, their health, and their needs.

Jane and I discussed this exciting research for a while, and she told me about the German forester Peter Wohlleben, who is also educating the world about the little-known secrets of the trees.

"It's interesting," she said, "that both Peter and Suzanne started off as foresters—managing forests so that they could be harvested in the most profitable way. Peter quit his job after fifteen years when he found that a forest he loved was doing quite well on its own, with almost no managing. And so he decided to devote himself to protecting—and understanding—that forest. He wrote a book: *The Hidden Life of Trees*. And I honestly think that this book has done for trees what *In the Shadow of Man* did for chimpanzees."

"Yes, and Suzanne has now written a book called *Finding the Mother Tree* that is having a similar impact," I said.

Jane was looking outside at the branches of a tree just beyond the window, faintly illuminated by the lights in our cabin. I wondered aloud what she was thinking.

"I am feeling wonder and awe about this incredible world we live in. And the truth is, we're destroying it before we've even finished learning about it. We think we are smarter than nature, but we are not. Our human intellect is amazing, but we must be humble and recognize that there is an even greater intelligence in nature."

"Do you have hope that we can find our way back to the wisdom of nature?" I asked.

"Yes, I do, but again, without head and heart working together, without cleverness and compassion, the future is very grim. But hope is essential, for without it, we become apathetic, and we will continue to destroy our children's future."

"Can we really heal all that we have harmed?"

"We must!" Jane exclaimed vehemently. "We've already made a start, and nature is there, waiting to move in and help to heal herself. Nature is so extraordinarily resilient. And remember, nature is so much more intelligent than we are!"

It was the perfect segue into Jane's second reason for hope.

REASON 2: THE RESILIENCE OF NATURE

"Let's take a walk," Jane said the next morning. We pulled on our jackets and went outside, the air chilly as the north wind greeted us, blowing through the trees of the preserve. "We can make something hot when we get back," Jane encouraged as we pulled the door closed. "It's good to have at least one walk a day," said Jane, after a few steps. "Though I don't really like to go for a walk without a dog."

"Why is that?"

"A dog gives a walk a purpose."

"How?"

"Well, you are making someone else happy." I thought of the rescue dogs at Jane's house in Tanzania, and how she never seemed happier than when she was surrounded by all creatures great and small.

It was a beautiful walk around a small lake, and Jane served as field guide, pointing out the sights she had seen the day before. The trees were mostly leafless, the land in winter slumber.

After we had walked for about thirty minutes, the sun broke through the clouds, illuminating a large tree in the distance.

"Let's go as far as that tree in the sun," Jane said, "and then turn back."

I was happy to move toward any spot of warmth. The tree was tilted to the side after many years of being buffeted by the strong winds.

When we arrived, Jane rested her hand on the moss-covered trunk of a magnificent Turner's oak.

"Here's the tree I wanted to come and say hello to. . . . 'Hello, tree.'" The tree sheltered us from the wind as the sun fell across our faces.

"It's beautiful," I said, touching the spongy green moss that Jane was stroking affectionately. Jane had told me that as a child she had a deep bond with a beech tree in her garden. She used to climb up and read her Doctor Doolittle and Tarzan books, disappearing for hours into the tree's leafy embrace and feeling closer to the birds and the sky.

"Did you have a name for that tree?"

"Just Beech," Jane said. "I loved him so much that I convinced my grandmother—we called her Danny—to give him to me for my fourteenth birthday and even drew up a last will and testament for her to sign, gifting me Beech. I used a basket and a long piece of string to haul up my books and sometimes even did my home-work up there. And I dreamed of going to live with animals in wild places."

"I know you've mostly studied animals, but you also learned a lot about plants when you were researching your last book, *Seeds of Hope*."

"Yes, and I absolutely loved that experience—what a fascinating world, the plant kingdom. And when you think about it, without flora there would be no fauna, would there! There would be no humans. All animal life ultimately depends on plants if you think about it. It's a kind of amazing tapestry of life, where each little stitch is held in place by those around it. And we still have so much to learn— we're like babes in the wood when it comes to really understanding nature. We haven't even begun to learn about the myriad of forms of life in the soil beneath us. Just think—the roots of this tree are way down there, knowing so much that we don't—and taking the secrets up to the branches above us."

As Jane looked from the ground up toward the very top of the tree, I had a vivid picture of her up in Beech, being rocked by the wind. I also thought of how her hands had rippled in flight in Tanzania as she was describing the murmuration of starlings and how she had said a naturalist needed to have empathy, intuition, even love. I wanted to know what she had found in the deepest mysteries of the natural world, and why what she discovered there gave her a sense of peace and hope for the future that I desperately wanted to find.

"Jane, you say the resilience of nature gives you hope—why?"

Jane smiled as she looked at the large tree in front of us. Her hand was still resting on its mossy, gnarled, and ancient bark.

"I think that I can answer your question best with a story."

I'd noticed that Jane often answered questions with stories, and I mentioned that to her.

"Yes, I've found that stories reach the heart better than any facts or figures. People remember the message in a well-told story even if they don't remember all the details. Anyway, I want to answer your question with a story that began on that terrible day in 2001—9/11—and the collapse of the Twin Towers. I was in New York on that day when our world changed forever. I still can

(Top) The Survivor Tree being rescued from Ground Zero, massively injured. The woman in the hardhat is Rebecca Clough, who initially realized the tree was still alive. Rebecca was the first of many dedicated people instrumental in her rescue and survival. (MICHAEL BROWNE) (Bottom) As she is now, thriving at the 9/11 Memorial & Museum. (9/11 MEMORIAL & MUSEUM, PHOTOGRAPH BY AMY DREHER)

remember the disbelief, the fear, the confusion as the city went quiet save for the wailing of the sirens on the police cars and ambulances on the streets emptied of people."

My memory flashed back to that brutal day as those two pillars of our modern world came crashing down. Having grown up in New York, I felt the attacks in a deep and personal way, as everyone knew friends or family who were in the towers or nearby when the attack occurred. I thought of the enormous crater at Ground Zero, the destruction, the horror of it all.

Jane continued her story. "It was ten years after that terrible day that I was introduced to The Survivor Tree, a Callery pear tree, who had been discovered by a cleanup worker a month after the collapse of the towers, crushed between two blocks of cement. All that was left was half a trunk that had been charred black with roots that were broken; and there was only one living branch.

"She was almost sent to the dump, but the young woman who found her, Rebecca Clough, begged that the tree might be given a chance. And so she went to be cared for in a nursery in the Bronx. Bringing that seriously damaged tree back to health was not an easy task, and it was touch-and-go for a while. But eventually she made it. And once she was strong enough, she was returned to be planted in what is now the 9/11 Memorial & Museum. In the spring her branches are bright with blossoms. People know her story now. I've seen them looking at her and wiping away tears. She truly is a symbol of the resilience of nature—and a reminder of all that was lost on that terrible day twenty years ago."

Jane and I stood quietly, thinking about the resilience of that tree. And then Jane began to speak.

"There's another story about a survivor tree which is even

more dramatic in a way," Jane said. "In 1990 I visited Nagasaki, the city where the second atomic bomb was dropped at the end of World War Two. My hosts showed me photos of the absolute and horrifying devastation of the city. The fireball produced by the nuclear explosion reached temperatures equivalent to the sun—millions of degrees. It was like a lunar landscape or what I imagine Dante's Inferno might look like. Scientists predicted that nothing would grow for decades. But, amazingly, two five-hundred-year-old camphor trees had survived. Only the lower half of their trunks remained, and from that most of the branches had been torn off. Not a single leaf remained on the mutilated trees. But they were alive.

"I was taken to see one of the survivors. It's now a large tree but its thick trunk has cracks and fissures, and you can see it's all black inside. But every spring that tree puts out new leaves. Many Japanese regard it as a holy monument to peace and survival; and prayers, written in tiny kanji characters on parchment, had been hung from the branches in memory of all those who died. I stood there, humbled by the devastation we humans can cause and the unbelievable resilience of nature."

Jane's voice was full of awe and I could tell she was far away, remembering that encounter.

I was moved by those two stories. But still I was not sure how the stories of these irrepressible trees could serve as one of Jane's main reasons that there was hope for our world and for our planet.

"But tell me, what does the survival of those trees tell you about the resilience of nature more generally?"

"Well, I remember a really bad bush fire at Gombe that swept through the open woodlands above the forested valleys. Everything

The tree that survived the atomic bomb that devastated Nagasaki in Japan. The great black wounds in her trunk show how she suffered. She is still alive and considered a sacred being by many in Japan. (MEGHAN DEUTSCHER)

was charred and black. Yet within a couple of days after one little shower of rain, the whole area was carpeted with the palest of green as new grass pushed up through the black soil. And a little later, when the rainy season began in earnest, several trees that I had been sure were completely dead started thrusting out new leaves. A hillside resurrected from the dead. And, of course, we see this resilience all around the world. But it's not just the flora; the fauna can regenerate, too. Think about the skinks."

"Skinks?" I asked.

"It's a type of lizard that shakes off its tail to distract predators who pounce on the wildly wiggling tail while the skink makes a getaway. Then, right away, the skink starts growing a new tail from the bloody stump. Even the spinal cord grows back. Salamanders grow new tails in the same way, and octopuses and starfish grow new arms. The starfish can even store nutrients in a severed arm, which sustain it while it grows back a new body and mouth!"

"But aren't we pushing nature to the breaking point? Isn't there a point at which resilience becomes impossible, a point where the damage suffered is irreparable?" I asked Jane. I was thinking of our emissions of greenhouse gases that trap the heat of the sun and that have caused temperatures around the globe to rise already by 1.5 degrees Celsius. And this, in addition to habitat destruction, is contributing to the horrifying loss of biodiversity. A 2019 study published by the United Nations reports that species are going extinct tens to hundreds of times faster than would be natural and that a million species of animals and plants could become extinct in the next few decades as a result of human activity. We've already wiped out 60 percent of all mammals, birds, fish, and reptiles—scientists are calling it the "sixth great extinction."

I shared these fears with Jane.

"It's true," she acknowledged. "There are indeed a lot of situations when nature seems to have been pushed to the breaking point by our destructive behavior."

"Yet," I said, "you still say you have hope in nature's resilience. Honestly, the studies and projections about the future of our planet are so grim. Is it really possible for nature to survive this onslaught of human devastation?"

"Actually, Doug, this is exactly why writing this book is so important. I meet so many people, including those who have worked to protect nature, who have lost all hope. They see places they have loved destroyed, projects they have worked on fail, efforts to save an area of wildlife overturned because governments and businesses put short-term gain, immediate profit, before protecting the environment for future generations. And because of all this there are more and more people of all ages who are feeling anxious and sometimes deeply depressed because of what they know is happening."

"There's a term for it," I said. "Eco-grief."

Eco-Grief

"I read a report by the American Psychological Association," I continued, "that found that the climate crisis can cause people to experience a whole range of feelings including helplessness, depression, fear, fatalism, resignation, and what they are now calling eco-grief or eco-anxiety."

"Fear, sadness, and anger are all very natural reactions to the reality of what is happening," Jane said. "And any discussion of hope would be incomplete without admitting the horrible harm we have inflicted on the natural world and addressing the real pain and suffering people are feeling as they witness the enormous losses that are occurring."

"Do you ever experience eco-grief?" I asked Jane.

"Quite often, and sometimes perhaps more intensely than others. I remember one spring day about ten years ago I stood with Inuit elders by the great ice cliff in Greenland and watched as water cascaded down and icebergs calved. The Inuit elders said that when they were young the ice there never melted, even in the summer. Yet it was late winter. They were weeping. That was when the reality of climate change hit me viscerally. And I felt pain in my heart for the plight of the polar bears as I watched rafts of ice floating where the ice sheet should have been firm and hard."

Jane's face was grim as she remembered the experience. "I flew from there to Panama," she continued, "where I met some of the indigenous people who'd already been moved off their islands because the sea levels were rising from the melting ice and warming water. They'd had to leave because at high tide their homes were endangered. Those two experiences, so close together, made a profound impact on me."

"It affects us viscerally when we see the places we love forever changed or destroyed," I said.

"We've also had enormous wildfires rage across Australia, the Amazon, the American West, and even in the Arctic Circle," Jane said. "It's impossible not to grieve for the harm we have inflicted, the suffering of people and wildlife alike."

It was just nine months after this conversation that the worst wildfire season in modern history began in California and other parts of the world. Over ten thousand fires burned four million acres or 4 percent of the entire state of California. One wildfire came within ten miles of where I now live in Santa Cruz. Almost a thousand families lost their homes in our area alone. For weeks on end the air was unbreathable; and on one particularly apocalyptic day, the sky remained dark and the sun never managed to break through the particulate-polluted sky. Touring the forests after the fire was like visiting an ash-covered, gray moonscape.

"I've spoken to someone," I said, "who has great insight into how we can confront and heal our grief."

I told Jane about Ashlee Cunsolo, who works with Inuit communities in Nunatsiavut, Labrador, Canada, who have been impacted by climate change. She was interviewing the communities about all they were losing—the ice that was breaking up; the temperatures that were rising; the plants and animals that were changing; and in many ways, an entire way of life that was disappearing.

"Cunsolo was hearing all these stories of despair and trying to write them up in her dissertation when she began experiencing radiating nerve pain in her arms and hands. The pain was so severe that she couldn't type or work.

"She went to all the medical specialists, but they could not find anything wrong with her nerves. Finally, she went to one of the Inuit elders and he told her, 'You're not letting go of your grief. Your body is stopping you from typing because you're intellectualizing it, not feeling it. Until you get it out of your body, your body won't function.' He told her she had to make space for her grief and speak it. And she also had to find awe and joy every day."

"What did she do?" Jane asked.

"She went into the forests. She immersed her hands in an ice-cold river and asked the water to take away the pain. She apologized to the land for the harm that she and others were doing. It was a reckoning.

"Cunsolo told me that she *had* been able to find awe and joy in the forest," I continued. "She said there's always beauty, even when there's pain and suffering. She learned not to hide from the darkness, just not to get lost in it."

"Did it help?" Jane asked.

"After two weeks of crying and letting the grief flow out of her body, the nerve pain was gone."

"That's an extraordinary and inspiring story. It speaks to something I feel deeply inside me," Jane said. "I've known a number of people who have been cured by indigenous healers, shamans, medicine men. And I have felt their power myself."

"Tell me more," I said.

"My first Native American friend—we call each other spirit brother and sister—is Terrance Brown, whom I know by his Karuk name, Chitcus. He inherited the role of medicine man of the Karuk tribe in California from his mother. One time I visited him when I was recovering from some unknown illness and feeling weak and somewhat depressed as I struggled to keep to my schedule. Chitcus got out his blanket in which he carries his drum and necklace of shells, a fan of eagle feathers, and a root of their sacred plant, which he spelled out for me as Kish'wuf. He lit the root until it gave off a sweet-smelling smoke, placed it in an abalone shell, and then, drumming softly, he chanted a healing prayer after which he took up the Kish'wuf and with the feathers gently brushed the smoke over my whole body as I stood with my eyes closed. My fatigue was gone after that.

"Since then he makes Kish'wuf smoke and prays for me every morning at dawn. He told me if the smoke rises straight up he knows I'm okay. Two of my other Native American friends—Mac Hall and Forrest Kutch—also pray for me every morning. No wonder I am still so healthy!"

"That's wonderful," I said, "and I think it speaks to the power of our interconnection—how an aspect of our healing lies in the quality of our relationships and the ways we come together to support each other."

Social support was certainly, according to the research, essential

Chitcus, my Native American "spirit brother,"
softly drumming as he very quietly chants a
prayer, then brushing the smoke from the
Kish'wuf that he holds in his left hand. (DR.
ROGER MINKOW)

for maintaining hope. Jane's words also reminded me of another
piece of Ashlee Cunsolo's story.

"Shortly after her healing, Cunsolo worked with five Inuit communities to make a film about their grief and loss," I said. "This
brought private pain into the light, and people started coming together to talk about how to heal and what to do next."

"They got together and expressed their grief," Jane said, "and that
helped activate them."

"Yes," I said. "Her story helped me see that facing our grief is
essential to combatting and overcoming our despair and powerless-
ness. The elders taught her that grief is not something to avoid or to
be afraid of. And that if we come together and share our sadness, it
can be healing."

"I absolutely agree," Jane said. "It's really important for us to confront our grief and get over our feelings of helplessness and hopelessness—our very survival depends on it. And it is certainly true—for me anyway—that we can find healing in nature."

"The trouble is that not enough people are taking action," I said. "You say more people are aware of the problems we face—so why aren't more trying to do something about it?"

"It's mostly because people are so overwhelmed by the magnitude of our folly that they feel helpless," she replied. "They sink into apathy and despair, lose hope, and so do nothing. We must find ways to help people understand that each one of us has a role to play, no matter how small. Every day we make some impact on the planet. And the cumulative effect of millions of small ethical actions will truly make a difference. That's the message I take around the world."

"But sometimes don't you feel that the problems are so huge that you are too overwhelmed to act, or else feel that anything you do is insignificant in the face of such enormous hurdles?"

"Oh, Doug, I'm not immune to all that's going on and sometimes it hits me. When, for instance, I return to an area that I remember as a peaceful patch of woodland with trees and birds singing—and find that in just two years it's been razed to the ground for yet another shopping mall. Of *course* I feel sad. But I feel angry, too, and try to pull myself together. I think of all the places that are still wild and beautiful and that the fight to save them must intensify. And I think of the places that *have* been saved by community action. Those are the stories that people need to hear; stories of the successful battles, the people who succeed because they won't give up. The people who, if they lose one battle, gear up for the next."

"Can these community actions win the overall battle, though?" I asked. "So many species are being lost. So many habitats destroyed—seemingly beyond repair. Is it not too late for us to prevent a total collapse of the natural world?"

Jane's eyes looked into mine, and her gaze was level and direct.

"Doug, I honestly believe we can turn things around. But—yes, there is a 'but'—we must get together and act now. We only have a small window of opportunity—a window that is closing all the time. So each of us must do what we can to start healing the harms we have inflicted and do our bit to slow down biodiversity loss and climate change. I have seen or heard about hundreds of successful campaigns and met so many wonderful people. And sharing these stories gives people hope—hope that we can do better."

Jane and I were meeting less than a month before the first cases of COVID-19 would be reported in China and a few months before public events would largely stop because of the pandemic. But we could not possibly have imagined any of that as we talked in the cabin in the Dutch forest. At the time Jane was still traveling nonstop, sharing her stories of hope around the world—often going to refugee camps and areas of extreme poverty, trying to comfort and uplift people in their darkest times of despair. I could only imagine the toll it must have taken on Jane.

"How do you keep your own spirits and energy up while trying to lift everyone else's?"

Jane smiled, and I could see the determination back in her eyes.

"When I travel and speak to people all over the world, the feedback I get is so heartening. People really want to believe that they can make a difference, but sometimes they need to hear from someone who has seen firsthand what people are doing. Seeing how people respond helps, but there's something else," Jane said, as she closed her eyes and took a deep breath. "When I was spending hours alone in the forest at Gombe, I felt part of the natural world, closely connected with a Great Spiritual Power. And that power is with me at all times, a force I can turn to for courage and strength. And sharing that power with others helps me to give people hope."

The sun had gone back behind the clouds. I wanted to hear more, but we were both shivering. "Shall we head back?" I suggested.

When we returned to the house, we quickly started a fire and as we sat down in front of it to eat a simple lunch, I urged Jane to tell more stories about the extraordinary resilience of nature.

"Well, first, you need to know that there are different kinds of resilience," Jane explained.

The Will to Live

"There's a kind of *built-in* resilience—as when spring brings forth leaves after a bitter winter of snow and ice, or the desert blooms after even a tiny amount of rain falls. And there are seeds that can germinate after lying dormant for many years. They contain that tiny spark of life just waiting for the right conditions to release its power. It's what Albert Schweitzer—one of my heroes—called the will to live."

"So life itself has an innate ability to survive and thrive?"

"Yes, absolutely. One of my favorite stories is about a grove of trees whose location is top secret. David Noble, a park ranger in Australia, discovered an unexplored and wild canyon. He rappelled down beside a waterfall and as he walked along the forest floor, he came upon some trees he was unfamiliar with. He picked a few leaves and subsequently gave them to some botanists for identification. They couldn't identify them at first, but imagine their excitement when they discovered that these leaves were identical with the fossilized imprint of a leaf found on an ancient rock. It was from a species long thought to be extinct—a species known only from the fossil record, a species that turned out to have survived for two hundred million years. Those trees, who came to be known as Wollemi pines, had been in that canyon, getting on with their lives, through seventeen Ice Ages!"

"What does that longevity say to you about resilience?"

"It merely says what many people are saying—that we need nature, but nature does not need us. If we restore an ecosystem in ten years, we feel we have made a big success. If it takes fifty years, it is hard to feel hopeful—the time scale just seems too long, and we are impatient. But it helps if we believe that in the end, even though we probably won't be around, nature will deal with the destruction we have caused."

"So you're saying nature plays the long game," I said, as I poured us both a coffee.

"Yes, and one thing I find really exciting is the absolutely amazing tenacity of life in seeds. After all the forest around Gombe was clear-cut, we started some tree planting—but it was really hard on the steep slopes. And we found that it was unnecessary, too, because the seeds of some trees, which must have been around for twenty years or so, began to germinate when the land was left to lie fallow. Even some of the roots of trees that had been cut down began to grow again."

She said there are many examples of this kind of spontaneous regeneration.

"My very favorite example of this is the story of Methuselah and Hannah," she said, "two very special date palms. Methuselah was the first to be brought back to life—from one of a number of seeds discovered in King Herod's desert fortress on the shores of the Dead Sea in the Jordan Rift Valley. Carbon dating revealed that these seeds were two thousand years old! Dr. Sarah Sallon, the director of the Borick Natural Medicine Research Center at Hadassah University Hospital, and Dr. Elaine Solowey, who runs the Center for Sustainable Agriculture at the Arava Institute for Environmental Studies at Kibbutz Ketura, got permission to try to germinate a few of them. One of the seeds grew—a male whom she called Methuselah after the character in the Bible, Noah's grandfather, who was said to have lived to be nine hundred and sixty-nine years old. When I met Sarah to learn more, she told me that she had been allowed to try to wake a few more of the precious seeds from their centuries-long sleep in the hope that one would be a female. That is how another ancient date palm, Hannah, began to grow.

"Recently I got an email from Sarah telling me that Methuselah had turned out to be a splendid partner, fertilized Hannah, and she had produced dates. Huge, luscious dates. Sarah sent me one—it arrived in a little cloth bag—and I was one of the first people to taste a date from the reincarnation of two Judean date palms from the long-gone forests of forty-foot palms that once grew throughout the Jordan Valley. The taste was utterly fantastic."

Jane closed her eyes and smacked her lips, recalling the sweet flavor of that resurrected date.

"Of course," she continued, "many species of animal have an amazing tenacity—a really strong will to live. The coyote that continues to spread across the United States despite the persecution from hunters. And rats and cockroaches—"

"I'm not sure it makes me feel more hopeful knowing that rats and cockroaches are going to survive long after we are gone!" I said.

Methuselah, above, grown from a two-thousand-year-old seed. Sarah Sallon, who worked to coax him from his long rest, was finally able to germinate a female seed, Hannah. Between them they produced mouth-watering dates. (DR. SARAH SALLON)

"Well, cockroaches are one of the most resilient and adaptable species."

"I certainly know that, having grown up in a big city. Cockroaches and rats were our wildlife. And pigeons, too."

"I know lots of people hate those species, but in fact, when they live out in nature, they are just part of the tapestry of life, each with a role to play. But just like us, they can take advantage of opportunities. They thrive on the food we waste and seize the opportunity for good living among the garbage so often found around human dwellings."

I wanted to find hope in Jane's stories of resilience, but I was still troubled. "So nature is extremely vibrant and strong, and it can adapt to the natural cycles that take place on the planet, but can it recover from all the harm we are doing to it?"

"Yes, I truly believe nature has a fantastic ability to restore itself after being destroyed, whether by us or by natural disasters. Sometimes it restores itself slowly, over time. But now, because of the terrible harm we are causing on a daily basis, we often need to step in and help in the restoration."

"So, Jane, I'm hearing that life is fundamentally resilient and can withstand great adversity. Are there any special qualities that we can learn from the resilience of nature?"

Adapt or Perish

Jane pondered for a bit. "Well, one really important quality of resilience is *adaptability*—all successful forms of life have adapted to their environment," she said. "The species that have been unable to adapt are the ones that have failed to make it in the evolutionary sweepstakes. It's our extraordinary success in adapting to different environments that has allowed humanity—and cockroaches and rats!—to spread across the world. So the big challenge faced by many species today is whether or not they can adapt to climate change and human encroachment into their habitats."

"Interesting," I said. "Tell me more about why some species adapt and others perish."

"For some species," Jane replied, "the life cycle, or specialized dietary needs and so on, is so preprogrammed that they may not be able to survive change. Other species are more flexible. It's fascinating to see how a species can survive if one or a few individuals manage to change—and pass on their behaviors to others in the group. While some individuals will die, the species as a whole will survive. Think of the plants that become resistant to the herbicides sprayed on the land by industrial agriculture and the bacteria that become resistant to antibiotics and end up as superbugs.

"But the stories I really love to tell," Jane continued, "are about highly intelligent animals who pass on information through observation and learning. Chimpanzees, for example, are a wonderful example of how a species can learn to adapt to different environments in a generation."

"In what way?" I asked, always eager to hear Jane's chimpanzee stories.

"The chimps in Gombe make a nest and go to bed at night, which is what most chimpanzees do. But the chimps in Senegal, where the temperatures are soaring and getting ever hotter, have adapted. They often forage on moonlit nights because it's much cooler. And they will even spend time in caves, which are very unchimp-like habitats.

"And chimps in Uganda have learned to forage at night for a different reason. Their forest is being increasingly encroached upon as villages expand and people need more land for farming. And so, with their traditional foods becoming scarcer, the chimps have learned to raid farms adjacent to the forest and make off with farmers' crops. That in itself is remarkable, as chimpanzees are typically very conservative in their habits and almost never experiment with new foods at Gombe. And if an infant tries to do so, the mother or elder sibling will hit it away! But the Ugandan chimps have not only developed a taste for foods such as sugarcane, bananas, mangoes, and papaya, but they've learned to make their raids in the moonlight—when they are less likely to encounter humans.

"But if we want an example of a really adaptable primate—in addition to ourselves, of course—we have to turn to baboons. They are quick to try any new food and, as a result, are extremely successful as a species, occupying many different habitats. And in Asia various species of macaques are also extremely adaptable. And, of course, because of their appreciation of human foods, they are considered pests and persecuted by us."

"So it sounds like adaptation is an essential part of resilience," I said, "and that some species and not others manage to adapt to a variety of new situations." I wondered if we would be able to adapt, not just to climate change but also to new ways of living that might slow it down.

"Yes, that's how evolution has worked for thousands of years. Adapt or perish. The trouble is, we've messed things up so much that we often need to intervene to stop the destruction of a habitat or the extinction of a species. And this is where the human intellect plays an important role: many people are using their brains to work with and support nature's innate desire to survive. There are so many wonderful stories of extraordinary people helping nature restore itself."

Nurturing Mother Nature

Jane was getting animated and leaned forward in her chair. Using her hands for emphasis, she made the point that we needed to realize that even when a habitat seems utterly destroyed, nature, given time, can reclaim that place, step-by-step. She said the first signs of life would be the really tough pioneer species that would create an environment for other life-forms to move in.

"There are people who study the ways of Mother Nature and copy them when they are trying to restore a landscape that we have destroyed," she explained.

"A great example is the restoration of a disused quarry, a monstrous five-hundred-acre scar where almost nothing grew near the coast in Kenya. This devastation had been created by the Bamburi Cement company. Interestingly, though, this vast restoration project was initiated not by a group of conservationists but by Felix Mandl, the man whose company had caused the devastation.

"He tasked the company horticulturist, René Haller, to restore the ecosystem. At first it seemed impossible; after searching for days, Haller found only one or two struggling plants sheltering behind the few rocks that had not been crushed. That was all.

"From the very beginning, Haller looked to nature to guide him in his work. He first selected a pioneer tree species most suited to the arid saline conditions—the casuarina, which is widely used in restoration projects. The seedlings took root and began to grow with some fertilizer and the addition of microfungi from the root systems of established trees. The trouble was their needlelike leaves did not decompose on the unrelenting ground, which meant other plants could not start to colonize the area. But then the ever-observant Haller, always wanting to learn from nature's wisdom, noticed that some beautiful millipedes with shiny black bodies and bright red legs were eagerly chewing up the needles—and their droppings provided just the right substance to create humus. He had hundreds of them collected from the surrounding countryside. The layer of fertile humus enabled other plants to grow.

"After ten years the original trees had grown to thirty meters, and the soil layer was now thick enough to support over one hundred and eighty species of indigenous trees and other plants. Various birds, insects, and other animals began returning to the land, and eventually giraffes, zebras, and even hippopotamuses were introduced. Today it is known as Haller Park and is visited by people from around the world, and it serves as a model for other restoration projects.

"It really is a fabulous story, isn't it?" Jane concluded. "It's not just about healing the harm inflicted by industry but about a CEO, way ahead of the greening efforts of companies today, undertaking the restoration just because he believed it was the right thing to do. It's a great example that even if we have totally destroyed a place, if we give it time and maybe some help, nature will return."

I wondered what the world would look like if we started that healing work in all the areas we'd desecrated. I had read a report that found that almost all the ecosystems studied recovered within ten to fifty years, with oceans recovering more quickly and forests more slowly. "Are you excited by the movement to 'rewild' parts of the world?" I asked Jane.

"I think it is a wonderful movement, absolutely essential," Jane said. "With so many people on the planet—and the vast majority of animals being humans and our livestock, including our pets—we must set some areas aside for wildlife. And the thing is, this rewilding is really beginning to work!"

Jane told me how, all across Europe, NGOs, governments, and the general public have agreed to protect large areas of forest, woodland, moorland, and other habitats and link them by means of corridors of trees and plants that allow mammals to move from one area to another in safety, something necessary to prevent too much inbreeding. The NGO Rewilding Europe has involved ten different regions across Europe in an ambitious plan that is already protecting a variety of habitats, creating corridors, and protecting or restoring a whole variety of animal species.

Jane's eyes were lit up with enthusiasm as she talked about these efforts.

"What are some of the animals that are coming back?" I asked.

"Let's see," Jane said, starting to tick her fingers. "Well, there's the elk, and those gorgeously curl-horned ibex, the golden jackal, which is really just a small gray wolf. Oh, and the common wolves, too, which hasn't always been met with enthusiasm. The Eurasian beavers, the Iberian lynx—a stunning feline that's still the most endangered cat in the world. Even, in some countries, brown bears. A whole variety of bird species are beginning to flourish such as whooper swans, white-tailed eagles, and griffon and Egyptian vultures. Some of these animals have not been seen in the wild for several hundred years."

"You speak of these different species so swiftly and with such familiarity, it's as though you're naming your friends."

"Well," she said, "it's because it matters a lot to me. These are the stories I think about to counteract all the gloom and doom."

"It is inspiring," I said. "And who's leading the way in saving these animals? Is it conservation groups? NGOs? Everyday people? What's making the difference?"

"Often it's everyday people," Jane said. "Some farmers are joining the rewilding efforts and turning their land back to nature, especially if it was not very suitable for farming in the first place. And some of the programs are really far-reaching and have a lot of support."

I told Jane about the farm my father-in-law had owned in Illinois and how he had planted native grasses and delighted in the wild turkeys and other species that had returned to the land. I'll never forget him on his tractor, surveying his land, farming the native plants. But wild turkeys are one thing, and some of the rewilding plans are hoping for the return of predators like wolves and mountain lions.

"I imagine some people are not fans of rewilding and don't want to set the land aside for animals, especially carnivores?"

"No, of course not," Jane said. "It is the same in Africa as in America. Farmers are worried about the loss of livestock to predators; fishermen and hunters are concerned about the effect some animals will have on their 'sport.' But as more people realize that animals have a right to live and are sentient beings with personalities, minds, and emotions, there is increased public support for these programs. What's really exciting is that some of these species in the European plains were on the very brink of extinction. A few dedicated people have saved all manner of highly endangered species from joining the long list of vanished life-forms and given them another chance."

"What is your favorite story about bringing a species back from the brink?"

Rescued from the Brink

"This story involves three very special characters," Jane began. "Dr. Don Merton, an adventurous wildlife biologist, and a female and male Chatham Island black robin. I loved this story from the start, as the European robin—on all the Christmas cards—is one of my favorite

Don Merton with a Chatham Island black robin. Don's passion and ingenuity helped rescue this imperiled species from the brink of extinction.
(ROB CHAPPELL)

birds, and the black robin looks the same, except for the color. The
two special birds were named Blue and Yellow for the colors of the
identification bands on their legs.

"I was able to meet Don during a tour in New Zealand, so I heard
it from the horse's mouth, as it were. Don is one of those truly inspirational people who gives me hope for our future. He was determined to save the last of these little birds from extinction.

"The problem is that there are no natural predators in New
Zealand, so when humans introduced cats, rats, and stoats, the
birds were easy prey as they had no built-in antipredator response.
They couldn't adapt to that kind of threat. Don wanted to help
save the species from imminent extinction, but he realized that
this meant he had to somehow catch the remaining black robins

and release them on an offshore island free of the introduced predators.

"By the time he got permission, and when the weather permitted him to check on them in the spring, there were only seven left. Only *seven* of these birds left on the planet. There were two females; and that first season, though both laid eggs, none of them hatched. These birds typically pair for life, but clearly their mates were infertile. Amazingly, though, for some miraculous reason Blue suddenly dumped her mate. She bonded with one of the three young males, they made a nest, and Blue laid the normal clutch of two eggs.

"Don told me he was in a terrible quandary. He had been involved in a successful breeding program of birds in captivity, but it involved a tricky, experimental method that also seemed rather hard on the parents—especially the mother. Did he dare take those two precious eggs away from Blue and put them in a tomtit's nest—a little bird about the same size as the robin—in the hope that Blue and Yellow would make another nest and produce two more eggs? He told me he felt terrible as he took Blue's eggs and destroyed that carefully made nest. The destiny of an entire species depended on whether the pair would nest again. And he would be responsible if they did not and went extinct.

"You can imagine the relief when they did make another nest and Blue laid two more eggs. And Don decided to do the same again. Another tomtit pair got two eggs to foster, and Blue and Yellow made a third nest and laid two more eggs."

I tried to imagine how they would have snuck the eggs out from Blue and Yellow's nest and slipped them into the other bird's nest without being noticed. "How did they get the tomtits to foster the eggs?" I asked.

"Well, cuckoos do it to all manner of birds. Birds often foster other birds' eggs. The real challenge happened after the first two eggs were successfully hatched. Don couldn't just leave the baby black robins to be raised by tomtits—then they would not learn black robin behavior. He put the minute chicks into Blue and Yellow's nest—and Yellow began feeding them. When Don took the next tiny chicks to add to the parental burden of Blue and Yellow, their third lot of eggs had hatched. The pair now had six chicks to feed instead of the usual two.

"Don told me that when he gently added the last two freshly hatched chicks into Blue's nest, she looked up at him as if to say, 'Whatever next?' but he told her, 'It's okay, darling, we'll help you with the feeding.' They gathered insects, grubs, and worms to help them feed their burgeoning family.

"Don and his team repeated the whole thing for the next few years, and most of the chicks fledged, mated, and produced chicks of their own. There are today about two hundred and fifty black robins.

"Just think—Don, Blue, and Yellow saved a whole species," said Jane. "Blue lived four years longer than the usual age. When she died at thirteen years old, she was famous and affectionately known as Old Blue. A statue was erected to her memory."

Jane clearly loved animal rescue stories and seemingly had countless others to share. She told me about so many more species that have been saved from extinction by human ingenuity and determination—mostly through captive breeding. The black-footed ferret of the great North American prairies was once thought to be gone forever until a farmer's dog killed one, and a search revealed a small population that had survived and which enabled scientists to start a successful captive breeding program. The whooping crane, the peregrine falcon, the Iberian lynx, and the California condor were all down to only

A female scimitar-horned oryx, after being reintroduced to her original wild habitat in Chad, gave birth to the first calf of this ambitious project. I got tears in my eyes when I was sent this photo. (JUSTIN CHUVEN/ ENVIRONMENT-AGENCY ABU DHABI)

a few individuals in the wild when successful efforts were made to save them. Species that were completely extinct in the wild but kept alive through captive breeding programs and that have now been re-introduced to their homeland include the milu deer in China and the Arabian oryx in Arabia. And there are so many more fish, reptiles, amphibians, insects, and plants that have been saved from extinction by the hard work and dedication of people who cared.

"I just got an email today with some wonderful news about the beautiful scimitar-horned oryx," Jane told me. "They were once found throughout desert regions of North Africa and Arabia but were hunted to extinction in the wild, and the species was only saved by captive breeding programs.

"I've been closely following the story of these stunning animals. The first twenty-five had been released into a huge area of their original habitat in Chad in 2016. Small groups have been released every year since, and now there are two hundred and sixty-five adults and adolescents and seventy-two calves, all roaming free and seemingly well adapted.

"This information came to me from Justin Chuven of the Environment Agency—Abu Dhabi. One question I asked him was whether it was really true that these oryx could survive for six months without drinking. He told me they often went without water for six to seven months and sometimes as long as nine months in a year!"

"That's an incredibly long time to go without water," I said. "How do they do it?"

"Justin told me the oryx are dependent on various water-rich plants, one of which is the very succulent but horrible-tasting bitter melon. He said he found it really entertaining to watch the oryx in a field of these fruits. They take one bite out of each melon, shake their heads in disgust, then move on to the next one and take a single bite from that—presumably hoping it will be less bitter! But it never is!"

I was inspired by these stories of heroic conservation, but I knew everyone did not believe that these rescue programs were worth the effort and the expense. "What," I asked, "do you say to people who think that these campaigns to protect endangered species are a waste of money? After all, throughout the history of life on Earth, ninety-nine point nine percent of all species have gone extinct, so people may wonder why spend the money to start saving species now?"

The Tapestry of Life

"Well, as you pointed out earlier, Doug, the rate of extinction today, due to human actions, is many, many times faster than ever before,"

Jane said, her face growing somber. "What we are trying to do is repair the damage we have created.

"And it's not just about benefiting animals. I try to make people understand how much we humans depend on the natural world for food, air, water, clothing—everything. But ecosystems must be *healthy* to provide for our needs. When I was in Gombe I learned, from my hours in the rain forest, how every species has a role to play, how everything is interconnected. Each time a species goes extinct it is as though a hole is torn in that wonderful tapestry of life. And as more holes appear, the ecosystem is weakened. In more and more places, the tapestry is so tattered that it is close to ecosystem collapse. And this is when it becomes important to try to put things right."

"Does this really work in the long term?" I asked, as we moved to be closer to the fire. I handed Jane the blanket and this time she drew it around her shoulders like a shawl. "Can you give me an example of the difference that these efforts can make?"

"I think the very best example is the restoration of the ecosystem of Yellowstone National Park in the United States."

Jane then explained how the gray wolf was wiped out a hundred years ago throughout most of North America. In Yellowstone, with the wolves gone, the elk overgrazed the park and the ecosystem suffered. Mice and rabbits could not hide, as the underbrush was gone, and their numbers plummeted. Bees had fewer flowers to pollinate. Even grizzly bears did not have enough berries to eat to prepare for their hibernation. The wolves had kept the elk away from the riverbank where they were exposed and vulnerable to attack. Without the wolves, the elk spent more time by the river, and the hooves of the large herds eroded the riverbank and caused the rivers to get muddy. Fish stocks declined in the cloudy water, and beavers could not build dams because overgrazing destroyed so many young trees.

Once reintroduced into the park, the wolves brought the elk population down from about seventeen thousand to a more sustainable four thousand. Scavenger species like coyotes, eagles, and ravens started thriving, as did the grizzlies. Even the elk were better off as their population stabilized at a healthier and more resilient size—they are no longer starving to death in the winter. For humans, the drinking water in the area around the park became cleaner and the tourism industry grew dramatically with the wolves back. I was beginning to see what Jane meant about the tapestry of life and the interconnection between all species.

"If only the media would give more space to the uplifting, hopeful news that we find everywhere," Jane concluded.

I asked Jane if she was ever asked if the money spent on conserving animals should be better used to help all the people in desperate need.

"You bet, I get asked that a lot," Jane said.

"How do you reply?"

"Well, I point out that I personally believe that animals have as much right to inhabit this planet as we do. But also that we are animals as well, and JGI, like many other conservation organizations today, does care about people. In fact, it has become increasingly clear that conservation efforts will not be successful and sustainable unless the local communities benefit in some way and become involved. They must go hand in hand."

"And you initiated this kind of program around Gombe," I said. "Can you tell me how that work began?"

"In 1987 I went to six countries in Africa where people were studying chimpanzees to find out more about why chimpanzee numbers were declining and what might be done about it. I learned a lot—about the destruction of forest habitats, the beginning of the bushmeat trade—that is, the *commercial* hunting of animals for food and the killing of mothers to sell their infants as pets or for entertainment. But during that same trip I began to realize also the plight faced by so many of the African people living in and around chimpanzee habitats. The terrible poverty, lack of health and education facilities, and degradation of the land.

"So I went on that trip to find out about chimp problems and realized they were inextricably linked to people problems. Unless we helped people, we could not help chimps. I began by learning more about the situation in the villages around Gombe."

Jane told me that she thought most people found it hard to believe the level of poverty that existed then. There was no appropriate health care infrastructure, and no running water or electricity. Girls were expected to end their education after primary school to help in the house and on the farm and were married off as young as thirteen. Many older men had four wives and huge numbers of children.

"There was a primary school in each of the twelve villages around Gombe. The teachers had canes that were freely used, and much of the children's time was spent in sweeping the bare earth of the schoolyard. Some villages had a clinic, but there were few medical supplies.

"And so, in 1994, JGI began Tacare. At the time it was a very new approach to conservation. George Strunden, who was the mastermind behind the program, selected a little team of seven local Tanzanians who went into the twelve villages and asked them what JGI could do to help. They wanted to grow more food and have better clinics and schools—so that's where we started, working closely with Tanzanian government officials. We did not even talk about saving chimpanzees for the first few years.

"Because we started with local Tanzanians, the villagers came to trust us, and gradually we built a program that included tree planting and protecting water sources."

"I heard you also set up microcredit banks?"

"Yes, I think this has been one of the really successful things we did. It was kind of magic that soon after Tacare began, Dr. Muhammad Yunus—who won a Nobel Peace Prize in 2006 and is one of my heroes—invited me to Bangladesh and introduced me to some of the women who had been among the first to receive tiny loans from his Grameen Bank. Dr. Yunus started this lending program because the big banks refused to give out small loans. The women told me it was the first time they had actually held money in their hands and the difference it had made. And that now they could afford to send their children to school. I was immediately determined to introduce this program to Tacare.

"On one of my subsequent visits to Gombe, the first recipients of the microloans that Tacare helped them obtain were invited to come and talk about the small businesses they had started. They were almost all women. One young woman—only about seventeen years old—was very shy but so eager to tell me how her life had changed. She had taken out a tiny loan and started a tree nursery, selling saplings for the village reforestation program. She was so proud. She had paid back her first loan; her business was making money; she was able to hire a young woman to help her; and she had actually been able to plan when she would have her second child, thanks to Tacare's family planning information. And she told us she did not want more than

A woman who received a Tacare loan and has begun a tree nursery. (JANE GOODALL INSTITUTE/GEORGE STRUNDEN)

three children because she wanted to be able to afford to educate them properly."

"I know you see voluntarily curbing population growth," I said, "and increasing access to education—especially for girls—as one of the keys to solving our environmental crisis."

"Yes, absolutely essential. On a visit to another village," Jane continued, "I gave a talk at the primary school and met one of the girls who had been awarded a Tacare scholarship that would enable her to move on to secondary education. She was very shy but excited at the idea of going to a secondary school in town where she would be a boarder."

Laughing, Jane told me how, at the start of this program, which was specifically designed to enable girls to stay in school during and after puberty, she had learned of a major problem. The girls were not going to school during their periods because the school latrines were stinky holes in the ground with no privacy whatsoever. Nor did they have any sanitary towels.

"So we planned to introduce 'ventilator improved pit latrines.' In America I suppose you would say VIP bathrooms. In the UK we'd say VIP loos!" She laughed again. "So that year I was asking for the money to build one of these as my birthday present. I raised enough for five! When they were built, I went to one of the schools for an official opening. It was a splendid event—parents in their smartest clothes, a few government officials, and a lot of excited children.

"The building had a cement floor, five little cubicles with doors that latched for the girls and, separated from these by a wall, three for boys. They had not yet been used. With great ceremony I cut the ribbon—then was escorted into the girls' area by the headmistress and a photographer. I went into a cubical and, to do the thing properly, sat on the seat. But I didn't pull my trousers down," she ended with a mischievous grin.

"So you see," she added, "these girls are now empowered to elevate their lives out of poverty and now understand that without a thriving ecosystem their families cannot thrive.

"Almost all of these villages have a forest reserve that needs protecting—but by 1990 most of them had been severely degraded—for firewood, for charcoal, and clear-cutting for growing crops. As most of Tanzania's remaining chimpanzees live in these reserves, the situation was not very promising. But now everything has changed. Our Tacare program is now in one hundred and four villages throughout the range of Tanzania's two thousand or so wild chimpanzees.

"Last year I went to one of these villages and met Hassan, one of their two forest monitors who had learned how to use a smart phone. He was eager to take us into 'his' forest and show us how he

Hassan is one of the forest monitors, trained in a Tacare workshop to use a smartphone to record animal traps or, in this case, an illegally cut tree. He also records sightings of chimpanzees, pangolins, and other wild animals. (JANE GOODALL INSTITUTE/SHAWN SWEENEY)

used the phone to record where he had found an illegally cut tree and an animal trap. He pointed out where new trees were now growing. He told us that he was seeing more and more animals—three days before he had seen a pangolin on his way home in the evening. And most exciting of all, he had seen traces of chimpanzees—three sleeping nests and some feces."

"I'm so sorry that I was not able to join you in Gombe," I said, thinking of my abrupt return to America to be with my dad at the hospital and while he was in hospice.

"You did exactly what you needed to do. There will be another time.

"What you would have seen is so truly exciting," Jane continued. "The program—which is all about taking care of people so they are better able to care for their environments—is working.

Emmanuel Mtiti has led our Tacare program from the very first. Wise and a born leader, he was the perfect choice to convince village leaders to join our efforts. Here we look out over the huge area where Tacare now operates to help people, animals, and the environment. (RICHARD KOBURG)

"The villagers are now so eager to learn about agroforestry and permaculture, and the farmers grow trees among their crops for shade and to fix nitrogen in the soil. All the villages have tree planting projects, and the hills around Gombe are no longer bare. And best of all, the people understand that protecting the forest is not just for wildlife but for their own future, and so they have become our partners in conservation."

Jane told me that the Tacare method is now operating in the six other African countries where JGI is working; and as a result, the chimpanzees and their forests, along with the other wildlife, are being protected by the people who live there, in whose hands their future lies.

"I see what you're saying about the link between nature's resilience and human resilience," I said. "How addressing human injustices like poverty and gender oppression makes us better able to create hope for people and the environment. Our efforts to protect endangered species preserve biodiversity on the Earth—and when we protect all life, we inherently protect our own."

Jane smiled and nodded her head, like an elder who was passing on the secrets of life and survival. I was beginning to understand.

I checked the time. It was almost four.

"Gosh, it's almost dark," Jane said. "Winter. Let's poke up the fire and have one last discussion and a wee dram. I need it for my voice." Indeed, her voice was sounding a little strained.

Jane got out a bottle of Johnnie Walker similar to the one that I had given her in Tanzania and poured two generous tots.

We settled down again, and Jane raised her glass. "Here's to hope," she said. We clinked and drank.

Our Need for Nature

"The last thing I want to say," Jane continued, her voice sounding stronger now, the whisky clearly having done its job, "is that not only are we part of the natural world, not only do we depend on it—we actually *need* it. In protecting these ecosystems, in rewilding more and more parts of the world, we are protecting our own well-being. There's lots of research proving this—but it is something that is incredibly important for me. I need time in nature—even if it's just sitting under a tree or walking in these woods or hearing a bird's song—to give me peace of mind in a crazy world!

"When I am in a hotel and looking out over a city, I think, 'Underneath all this concrete, there's good earth. We could be growing

things. There could be trees and birds and flowers.' Then I think of
the push to green cities with urban trees, which not only reduces temperatures by several degrees, reduces air pollution, and improves water
quality but also improves our sense of well-being. Even in cities, like
Singapore, there are projects now that link small areas of habitat with
green corridors of trees so that animals can move from place to place
as they look for food and mates. Whenever you give her a chance,
nature returns. Every tree planted makes a difference."

I knew Jane had been involved in an initiative that was launched
at the Davos World Economic Forum to plant a trillion trees to counteract the global deforestation that humans have been responsible for.

"The trees may save us," I said.

"Planting trees is very important," Jane said. "Protecting forests
is even more important—it will take time for saplings to grow big
enough to absorb enough CO_2. And, of course, they must be looked
after. And, of course, we must clean up the ocean, too—and obviously we must reduce greenhouse gas emissions."

"Where do you go to restore yourself in nature when you are
not at Gombe?"

"Each year I try to go to Nebraska, to the cabin of my friend Tom
Mangelsen, who is a wildlife photographer. It is on the Platte River,
and I go during the migration of the sandhill cranes and snow geese
and many other species of waterbird."

"Why do you go there?" I asked, knowing that she could go any-
where in the world on her endless travels.

"Because it is a dramatic reminder of the resilience we have been
discussing. Because despite the fact that we have polluted the river,
despite the fact that the prairie has been converted for growing genetically modified corn, despite the fact that the irrigation is depleting

the great Ogallala Aquifer, despite the fact that most of the wetlands have been drained—the birds still come every year, in the millions, to fatten up on the grain left after the harvest. I just love to sit on the riverbank and watch the cranes fly in, wave after wave against a glorious sunset, to hear their ancient wild calls—it is something quite special. It reminds me of the power of nature. And as the red sun sinks below the trees on the opposite bank, a gray, feathered blanket gradually spreads over the whole surface of the shallow river as the birds land for the night, and their ancient voices are silenced. And we walk back to the cabin in the dark."

Jane's eyes were closed and her face was glowing, no doubt reliving and being renewed by recalling those magical evenings.

As I sipped my whisky, I felt the warmth in my chest. "I must tell you about an unforgettable experience I had in nature that gives me hope," I said.

"Tell me," Jane said, eager for another story to add to her collection.

"The Pacific gray whales that were almost hunted to extinction and are now not only bouncing back but also coming to interact with humans, their former mortal enemies. These whales are called the friendly gray whales."

"Yes, I've heard of them. It's quite amazing."

"I had an experience in the whale nursery in Baja, Mexico, that moved me deeply. I noticed that one whale was extremely white, which our guide explained occurs with these whales as they get older. Its body and tail had numerous scratches and gouges, which usually come from years of defending babies from orcas that try to eat the young on their annual migration from Alaska to Baja. As the whale came closer, we could see many barnacles on its skin and a deep indentation in the back of the blowhole, which also were signs of an elder whale. Our guide said it was almost certainly a grandmother whale.

"The grandmother whale's head popped up next to our boat as the swirling, bubbling water spilled away. She raised her chin toward the rail of our boat, and we began to stroke her silvery skin. Aside from the barnacles, her skin was smooth and spongy, as we could feel the soft blubber beneath. As we stroked her she rolled to her side, opening her mouth and showing us her baleen, a sign of relaxation. And then she looked at us with one of her beautiful eyes. What she could see of us as we stared down at her from the boat, smiling and laughing, I had no idea, but it was clear she felt safe and wanted to connect in these bays, where possibly during her lifetime we had almost exterminated her kind. I felt so moved that tears were rolling down my cheeks.

"Our guide was in the background saying, 'This whale has forgiven us. She has forgiven us for who we were and is seeing who we are today.'"

"It's extraordinary when we recognize our connection to the natural world," Jane said, nodding.

"Can you tell me more about the places where you feel this connection most strongly?" I encouraged.

"Well, of course I go to Gombe each year. I sit on the peak where I once sat as a young woman and look out over Lake Tanganyika toward the distant mountains of the Congo. And at the edge of that vast lake, the longest and second deepest in the world, the sun sinks below the mountains and the sky turns the palest pink and then crimson. Or the black rain clouds build and the thunder growls and the lightning flashes, and it is night.

"And there are times when I lie on my back in some quiet place and look up and up and up into the heavens as the stars gradually emerge from the fading of day's light. And I see myself, a tiny speck of consciousness in the enormity of the universe."

At that moment, I felt as though I could sit forever by the fire and listen to Jane's stories, but as I looked out the window at the first stars, I knew it was time to go, to rest, so we could return the next day to explore her last two reasons for hope. "Shall we stop for the night?" I asked.

"I just want to share one last story about hope and the resilience of nature," Jane said, coming out of her deep reverie.

"Last year, on the UN International Day of Peace, I took part in a very special ceremony in New York. There were about twenty members of JGI's international youth program, Roots & Shoots, many of them African American high school children from across America. We gathered around The Survivor Tree—the tree who was rescued after she was crushed and wounded on 9/11. The devoted nurseryman Richie Cabo, who had helped to heal her, was with us. We looked up at the strong branches reaching toward the sky.

"Only a short time before they had been filled with beautiful white blossoms, and now the leaves were beginning to fall. We stood silently and prayed for peace on Earth, for an end to racial hatred and discrimination, for a new respect for animals and nature. I looked around at the young faces, the faces of those who would inherit the planet wounded by countless generations of humans. And then I saw it, I saw the neat perfection of the nest of some small bird. I imagined the parents feeding the nestlings, the fledging, the final hopeful flight into the as yet unknown world. The children were also staring up at the nest. Some smiled, others had tears in their eyes. They, too, were

Visiting The Survivor Tree on the UN International Day of Peace. Only deep wounds in her trunk tell the story of her suffering. Two of the men who gave her the chance to survive are with me—Richie Carbo, the nurseryman, is closest to me, and on my far right is Ron Vega, who made sure she had a home at the memorial site. (MARK MAGLIO)

ready to move out into the world. And The Survivor Tree, brought back from the dead, had not only put out new leaves herself but nurtured the lives of others."

Jane turned to me, there in the little cabin in the woods in the Netherlands.

"Now do you understand how I dare hope?" she asked quietly.

REASON 3: THE POWER OF YOUNG PEOPLE

"I've always wanted to work with children," Jane said. "It's funny, when I was young, I used to think, one day I'll be old—as I am now—and I always pictured myself sitting on a rustic seat under a tree telling stories to little groups of children."

With Roots & Shoots group invited to the United Nations with me for UN International Day of Peace. (JANE GOODALL INSTITUTE/MARY LEWIS)

It was easy to imagine Jane under her beloved beech tree, surrounded by children. I could see the trees outside through the two windows beside us, but was grateful that we were cozy inside, sitting by the fire.

The morning sun was making Jane's cheeks glow as we began another day of interviews. Looking at her in her salmon-colored turtleneck and gray puffy jacket, I realized I never thought of her as being old. There was something so vibrant, so alive, so unstoppable about her. I marveled at how differently people age: some people in their forties and fifties seem like they have been defeated by life and begin to recede; some in their eighties and nineties seem to be endlessly curious and engaged with all that life's laboratory has to discover.

Just then, as if to prompt us in the direction of Jane's third reason for hope, we heard children laughing outside.

"The talks that I like giving best," Jane said, "are to high school and university students. They are so engaged, so alive. But it actually works better than you think even with the little ones. They're wriggling about on the floor, and you're telling them stories and thinking, 'Well, they're not really listening.' And then I meet the parents, and the children have told their parents exactly what I've said. They're not supposed to be sitting still at that age—same with a baby chimp—because they're learning and listening while they play. That's why school can be so bad. It keeps small children sitting. It's awful, they're not meant to do that. They're meant to be learning from experience. Fortunately, more schools are beginning to change now, taking children out into nature, answering their questions and encouraging them to draw and tell stories."

"How did you start working with young people?" I asked.

"As I began traveling around the world, raising awareness about the environmental crisis, I met young people on all continents who were apathetic and disengaged, or angry and sometimes violent, or deeply depressed. I began talking to them and they all said more or less the same thing: 'We feel this way because our future has been compromised, and there's nothing we can do about it.' Indeed, we have compromised their future.

"There is a famous saying," Jane continued. "'We have not inherited the Earth from our ancestors but borrowed it from our children.' And yet we have not *borrowed* it from our children. We've stolen it! When you borrow something, the expectation is that you will repay. We have been stealing their future for countless years and the magnitude of our theft has now reached absolutely unacceptable proportions."

"It's not just from this generation that we are stealing," I added. "We are stealing it from all future generations. Some are calling it intergenerational injustice because the children of the future, the people of the future, do not have a vote or a say in our boardrooms."

"Yes, it truly is that," Jane commented. "But I didn't agree with the young people who told me there was nothing they could do about it. I told them we have a window of time when, if people of all ages—young and old—get together, we can at least start healing some of the hurt we have inflicted and slow down climate change.

"If *everyone*," she continued, "starts to think about the consequences of what we do, for example, what we buy—and I am including young people thinking about what they ask their parents to buy for them—if we *all* start to ask whether its production harmed the environment, or hurt animals, or is cheap because of child slave labor or unfair wages—and, if so, we refuse to buy it—well, billions of these kinds of ethical choices will move us toward the kind of world we need."

This hopeful philosophy of "everyone can make a difference" led Jane, in 1991, to begin her youth program, Roots & Shoots.

"Can you tell me about how Roots & Shoots started?" I asked Jane.

"Twelve Tanzanian high school students from eight schools came to my home in Dar es Salaam. Some of them were worried about things like the destruction of coral reefs by illegal dynamiting, and poaching in the national parks—why wasn't the government clamping down on these? Others were concerned about the plight of street children, and others about the ill treatment of stray dogs and animals in the market. We discussed all this and I suggested they might do something to improve things.

"So they went back to their schools, gathered together others who were also concerned, we had another meeting, and Roots & Shoots was born. Its main message: every single individual matters, has a role to play, and makes an impact on the planet—*every single day*. And we have a choice as to what sort of impact we will make."

"It's not just about the environment, is it?"

"No. Understanding that everything is interrelated, we decided that each group would choose three projects to help make the world a better place—for people, for animals, and for the environment—starting in their local community. They would discuss what could be done, prepare themselves for the work ahead, then roll up their sleeves and take action."

"How did people respond to these students taking action?"

"The first Roots & Shoots group was mocked for cleaning up a beach without being paid—you only worked for nothing for your parents, because you had to!" Jane said, chuckling. "But soon an explosion of activity gave rise to a new phenomenon in Tanzania: volunteerism.

"It was a grassroots program, and gradually more schools became involved. Many groups chose to plant trees in their barren schoolyards; and within a few years—trees grow fast in the tropics— all those schools had shady areas where students could relax or work surrounded by trees and birds."

Roots & Shoots has since become a global movement, with hundreds of thousands of members, from kindergarten through university, active in sixty-eight countries—and growing.

"What gives me hope," Jane continued, "is that everywhere I go, young people filled with energy want to show me what they've done and what they're doing to make the world a better place. Once they understand the problems and when we empower them to take action, they almost always want to help. And their energy and enthusiasm and creativity are endless."

"The public perception is that many young people, especially privileged ones in developed countries, are materialistic or self-centered," I suggested.

Jane agreed that it was true in some cases, but certainly not always. "We have lots of Roots & Shoots programs in private schools; and children who come from privileged backgrounds are often deeply engaged and truly want to help. They just need a few stories to reach their hearts and wake them up to the satisfying feeling they get from doing something to help."

"That's certainly true for my own children," I said. "Over the years I've seen how their growing awareness of the world's problems has galvanized them to adopt causes that matter to them. I wonder what it's like for children who are themselves struggling. I know you've also worked with young people living in extreme poverty and in refugee camps."

"Yes, I have found children living in underprivileged communities are very motivated to help others. I am always very moved when I see the excitement in the eyes of these children when I tell them they can make a difference. That the world needs them. Above all, that they matter."

Jane paused and seemed lost in her thoughts. I waited for her to speak.

"I was just thinking about the first time when I knew for sure that the program was going to work," she said. "I knew it was doing really well in Tanzania and in an international school and a private school in America. But what about a low-income public school in the Bronx—any chance that we could help empower youth to make a difference there?"

Jane was introduced to a teacher, Renée Gunther, who arranged for Jane to come and speak at what she was told was the second poorest elementary school in the state of New York. "Almost all the children had older siblings or fathers who were in gangs," Jane said, "and both drug addiction and alcoholism were rife. In a run-down auditorium I talked to the children about the chimps and about Roots & Shoots. To my delight, many of them seemed really interested and there were a lot of questions, especially about the slide I had shown of a chimpanzee dressed up for the circus. When I explained how the training of chimps for circuses is very cruel, and how they have all been taken from their mothers, it was clear that these kids felt empathy with them."

The following year Renée asked Jane to pay another visit. "I met with her and the principal, and they told me that some of the children had been really keen to start Roots & Shoots groups and wanted to tell me what they had decided to do. 'I'm sure you've seen far more polished presentations,' the teacher told me afterward, 'but these children have never presented before.' There were actually tears in her eyes.

"The first group of children wanted to ban Styrofoam from their school lunches. They had developed a little skit," Jane recalled. "One boy acted the part of manager of a company and another the spokesperson for the small group of Roots & Shoots members. Not only were they amazingly knowledgeable about Styrofoam—they all acted really well! They were actually invited later to make their presentation in front of the borough president of the Bronx. And they succeeded in getting Styrofoam banned from their school!"

"That must have made these kids feel so proud," I said. "And made them feel that it really is possible for kids to make a real difference."

"Yes, that's what's so exciting," Jane agreed. "Then there was a presentation by Travis, an eleven-year-old African American boy, that impressed me even more. His teacher had told me that before he joined Roots & Shoots, he had seldom attended school. When he had, he'd sat at the back of the class, hiding his face under the hood of his sweatshirt. He never spoke.

"Well, Travis came forward and stood very straight in front of me, looking directly at me. The one other member of his group stood silent behind him. Travis told me he had seen a dressed-up chimpanzee on a cereal box. He was supposed to be smiling and happy. 'But you told us that face wasn't smiling, it was because he was fearful,' he said. 'So I wrote to you and you told me that I was right.' Standing even straighter, he looked right into my eyes. 'That's when I decided to *take action*,' he said. He and his friend had written to the manager of the company. They got a letter back that thanked them. Many others had objected to the chimpanzee on the cereal box, too, but Travis did not know that. Imagine how he felt when the advertisement was withdrawn!"

"One of the most important determinants of hope in one's life is seeing *one's agency*, one's ability to be effective," I commented. "That must have changed his life. It makes you wonder what small success started Gandhi or Mandela on their life paths."

"Yes, this is why I am so passionate about working with youth in all walks of life. So often they just need a chance, just need some attention, someone who listens and encourages and cares. If they have that support and begin to see that they can truly make a difference, then the impact they can make is enormous."

Love in a Hopeless Place

Jane told me many more inspiring stories about Roots & Shoots groups and how they were transforming their communities. I was particularly moved by the story she told about the encounter that fueled her desire to start Roots & Shoots on reservations in the United States.

"In 2005, after one of my talks in New York," Jane said, "a note was brought to the stage door. It was from a Native American man named Robert White Mountain, asking if he could come backstage and talk to me.

"I was stunned when he told me his sixteen-year-old son had recently killed himself by hanging."

Robert White Mountain told Jane that the place where he lived in North Dakota had one of the highest suicide rates in the United

States—there were three to six suicides or suicide attempts a week. Only fifteen of the young men his son had gone to school with were still alive. As he was burying his son, Robert silently promised him he would try to do something about it. He had heard about this woman called Jane Goodall and her Roots & Shoots program and, in his desperation, wondered if she could help.

"So—could you?" I asked Jane.

"Well, I did manage to get to his community. He took me to visit the safe house he had created for young people who were affected by drugs, alcohol, and violence in their homes. It was a tiny building with no windows and very little furniture; and there he told me about life on his reservation—the extremely high poverty and unemployment rates that often led to hopelessness, helplessness, alcohol, drugs, violence.

"It was unimaginable to me that a community like this, where people are living in conditions that are worse than those in many developing countries, could exist in the middle of the wealthiest country on this planet."

As Jane spoke, it was clear the memory of this conversation still upset her. She told me how Robert had said that his people used to be called the caretakers of the land but that over the years they had lost that connection.

"That encounter fifteen years ago led to many more meetings and friendships with some wonderful elders and chiefs throughout the United States and Canada," Jane said. "I have connected with a number of them on a deep spiritual level."

"Were you ever able to establish programs on any reservations in the United States?" I asked Jane.

"So far only on one," she said. "On the Pine Ridge Reservation in South Dakota, another community where alcohol and drugs and suicides are common. It began in an unexpected way. I had organized a gathering of tribal elders to meet with me and some of my staff in South Dakota to discuss how we might start such a youth program. I invited a young man, Jason Schoch, who I'd met a couple of years before during a period when he was deeply depressed. I knew he wanted to draw on his own experiences to reach out and help young people. In the end there was only a very small group of us—none of the local chiefs could attend because of a sudden, unseasonable snowstorm. But there was one young woman who showed up: Patricia Hammond, whose mother was Lakota. Patricia and her family lived on the Pine Ridge Reservation. Even though Patricia hadn't met Jason before, they spent the entire time we were snowbound planning how to start Roots & Shoots on the reservation. Jason returned to California, where he was working, but—well, when he could no longer afford his nightly calls to Patricia, he moved to South Dakota to join her!"

Jane told me that Patricia and Jason started by working out ways to reconnect the young people of Pine Ridge with nature and their culture. They got a group to help remove trash and start a small organic garden. They wanted to teach them about traditional foods and medicinal plants. "They revived the traditional Hidatsa or 'three sisters' plots," Jane explained, "planting corn, beans, and squash together." These plots produce plentiful high-quality yields with minimal environmental impact. The corn provides the support for the beans to climb; the beans replenish nutrients in soil; and the large squash leaves provide living mulch and serve to provide shade, conserve water, and control weeds.

"During that first season everything flourished on the little plot.

Patricia Hammond taught the Roots & Shoots groups in the Pine Ridge Reservation in South Dakota about traditional plants, working with the elders. (JASON SCHOCH)

The corn grew to be six feet tall, but just as the kids were getting excited to harvest it, one of the Roots & Shoots members had an especially difficult weekend—he had a meltdown. He broke through the fence and cut down and trampled all the corn.

"Patricia told me she felt like giving up," Jane said. "Instead, she and Jason and the Roots & Shoots kids mended the fence and started again. Patricia and Jason ultimately created twelve community gardens and three farmers markets for the community. She said the gardens helped her community reconnect with the land once again and feel hopeful and joyful.

"I think that the three pillars of Roots & Shoots," Jane concluded, "helping people, animals, and the environment—really tie in with the ancient belief of many indigenous people that we are all one."

When the youth began planning and helping with projects, those who joined the program gained a sense of purpose and self-worth that they had lacked. "Roots & Shoots really made a difference," Jane added. "Many members graduated from high school and a couple went on to university. And Jason and Patricia are still nurturing and expanding their work on the reservation."

"It is inspiring," I said, "that the program can make a difference even in a community where so many other programs have failed to help."

Jane smiled. "I think it's successful for a whole variety of reasons," she said. "Firstly, because the young people get to have a say in their

The children of Pine Ridge have taken to gardening with pride and delight. (JASON SCHOCH)

activities. It's a bottom-up movement. And if they *choose* a project, then they work at it with great enthusiasm and passion. Secondly, most of the groups are in schools, and all the teachers who agree to get involved do so because they are inspired by the concept—that it is a program that embraces all the concerns and interests of the different members of the group. There are always some students who want to help and learn about animals, some who are most concerned with social issues, and some who are passionate about their environment. In addition, it links young people from different countries and is a great way of learning about other cultures."

The more Jane spoke about these kids, the more animated she became.

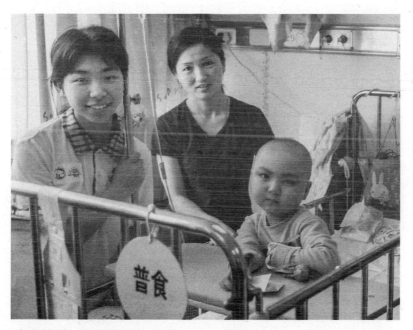

Chinese university Roots & Shoots students visiting a child with cancer, bringing toys and telling stories. (JANE GOODALL INSTITUTE / CHASE PICKERING)

"All these young people are changing the world for the better every day," she continued, "and every time they accomplish their project successfully, their feeling of empowerment increases, and they become more self-assured. And because we are always looking to partner with other youth organizations that share our values, the students become more and more hopeful that together they will succeed. And again and again, they achieve the hoped-for outcome."

Jane's stories affirmed that when we feel we can make a difference, and we're given the means to do so, positive outcomes can happen that in turn allow hope to prevail. It was a powerful example of what the research had found contributes to hope: clear and inspiring goals, realistic ways to realize those goals, a belief that one can achieve those goals, and the social support to continue in the face of adversity.

Jane told me another story of hope in a desperate place, a camp for Congolese refugees run by the United Nations High Commissioner for Refugees—UNHCR—in Tanzania. She explained that Roots & Shoots was first introduced into the huge camp by an Iranian staff member of UNHCR, but he left after a short time and three young Tanzanian volunteers, one after the other, had continued the task. "They coped with endless bureaucracy, were given a tiny bare space as an office and somewhere to live, and eventually got Roots & Shoots groups into several of the schools." She said they also organized and got funding for programs to teach the members skills—like organic vegetable gardening, hairdressing, cooking, and raising chickens, which eventually successfully replaced illegal hunting for bushmeat.

"On one of my visits, each Roots & Shoots family was gifted a hen and a cockerel," Jane explained. "We knew the animals would be well looked after because their children had learned how to feed them and keep them safe at night. During the day they pecked around the houses. For both parents and children these gifts were seen as precious—they had so few possessions. And soon these hens would produce chicks, and they would get a little flock and have the luxury of adding eggs to their handouts of rice and cassava. Of course, our Roots & Shoots groups were providing fresh vegetables, too."

"What happened to the refugees?" I asked.

"Soon after that, they were forcibly repatriated to the Democratic Republic of the Congo. Many were frightened to return home because they had no families left there—they had all been killed in the horrible fighting. I was told that the Roots & Shoots groups took their chickens and the seeds they had saved from their vegetable gardens with them when they returned."

She told me that for the first few months, UNHCR housed the returning refugees in a reception camp while they tried to sort out their futures.

"About two months later," Jane continued, "I received a letter from a visitor to this camp. It was depressing, he said—bare earth, people with vacant expressions, children sitting listlessly outside their huts. He continued walking through the camp, and then suddenly he came to a section of the camp where the atmosphere changed. Children were running around and laughing. Hens foraged in a patch of land where grass had been allowed to grow. A few teenagers were working in a small vegetable garden. The visitor asked his host why it was different there. 'Well, I don't really know. But it's something called Roots & Shoots.'"

Children of Chinese migrants from country to city. Chinese university students helped them understand that they mattered and could make a difference. (JANE GOODALL INSTITUTE ROOTS & SHOOTS, BEIJING, CHINA)

"I Don't Want Your Hope"

Of course, Roots & Shoots is but one of a number of organizations working to empower, educate, and activate young people. All around the world young people are increasingly taking to the streets to demand change. "Fridays for Future" was initiated by Greta Thunberg, the environmental activist who at the age of fifteen started protesting outside the Swedish Parliament with a sign that read SCHOOL STRIKE FOR CLIMATE. Greta has spoken with world leaders and at major conferences, and millions of people have participated in these youth-led climate protests.

I asked Jane if she had met Greta Thunberg.

"I have. She's done an amazing job of raising awareness of the climate crisis in many parts of the world and not only among the youth."

I wondered what Jane thought of Thunberg's provocative speech at the World Economic Forum when she declared, "Adults keep saying: 'We owe it to young people to give them hope.' But I don't want your hope. I don't want you to be hopeful. I want you to panic. I want you to feel the fear I feel every day, and then I want you to act. I want you to act as you would in a crisis. I want you to act as if our house is on fire. Because it is." I asked Jane what she thought about Greta's critique of hope and her belief that fear is a more appropriate response.

"We *do* need to respond with fear and anger about what is happening," Jane replied. "Our house *is* on fire. But if we don't have hope that we can put the fire out, we will give up. It's not hope *or* fear—*or* anger. We need them all."

"We have so many huge problems. Isn't it a cop-out for adults to say that the children will solve these problems?"

Jane sat up in her chair, clearly provoked by my question.

"It actually makes me angry when people say it will be up to young people to solve them," she said. "Of course, we can't and shouldn't expect them to solve all our problems. We've got to support them, encourage them, empower them, listen to them, and educate them. And I truly believe the young people of today are rising to the challenge in a most remarkable way. Once they understand the problems and are empowered to take action—well, they are changing the world as we speak.

"And it's not only what *they* do," Jane added. "It's particularly exciting to see how children are influencing their parents and grandparents. So many parents tell me that they never thought about their purchases until their child started explaining what they were learning about the environment."

"How did that work?" I asked, thinking back to my own experiences as a parent and how my children became advocates for buying green and were the driving force behind many of the changes my family made in the way we shop and consume.

Jane elaborated. "One of the best examples I know comes from China. In 2008, a ten-year-old girl named Joy attended one of my talks, and afterward begged her parents to help her start the first Roots & Shoots group in Chengdu. Their motto was a quotation of mine: 'Only if we can understand, can we care. Only if we care, will we help. Only if we help, shall all be saved.' At first the children simply followed suggestions made by their teachers, but it was not long before they could design and conduct projects on their own. They became one of the most active groups. A few years later I received a letter from Joy's mother translated from the Chinese by her daughter—who had learned English in order to communicate with me! I have to read it to you." Jane sprang up and fetched her laptop.

"This is what she said: 'After our children formed a Roots & Shoots group at school, they changed how we thought. It is no exaggeration to say that most of us would never have thought to care for the environment without our kids. And we might still have the numb lifestyle, caring nothing about this planet but ourselves. Our kids used a bright way to let us have a different view of our life. I started my own change from accepting passively to actively participating after my child brought back all the information from Roots & Shoots. I went from a consumer who was quite selfish to one who learned to cut down unnecessary buying.'"

"What an amazing letter," I said, when Jane had finished reading it. I then learned that the story gets even better.

Jane has kept up the personal correspondence with Joy over the

All over the world young people are clearing
up litter—in the streets, on the beaches, and
installing recycling bins in school cafeterias,
such as here in Kibale, Uganda. (JANE GOODALL
INSTITUTE/MIE HORIUCHI)

years and so learned that Joy's mother became such an enthusiast
that she began to design courses and write plays related to environmental protection.

And Joy, now eighteen years old and attending university, has
rallied the local government behind her Roots & Shoots group in a
highly successful recycling program to make Chengdu free of litter.

Jane's story about Joy and her mother reminded me of a story
I heard from Christiana Figueres, one of the architects of the Paris
Agreement. At a gathering of the World Economic Forum, Ben van

Beurden, the CEO of Royal Dutch Shell, had requested to meet with her one-on-one, without the usual accompanying staff on either side. At the close of their meeting he said, "Christiana, let me be very frank. We are both parents." He told her about a profound moment when his ten-year-old daughter came to him and asked whether it was true that his company was destroying the planet. He made a pledge to her that he would do anything to make sure she would grow up on a planet that was safe and sustainable for her and future generations. And he decided to support the Paris Agreement.

Clearly, Jane and her thirty-one Jane Goodall Institutes around the world were inspiring a young army of environmental defenders in more than sixty-five countries, but I still wondered if it was enough to change the values of our extraction and consumption civilization without leadership from the top. I wanted to know exactly why she put so much hope in this next generation and whether it was misguided to think that young people could really address the problems previous generations had created.

Millions of Drops Make an Ocean

"I have heard from many of the older visionaries I have met that young people give them hope," I said, "and I am still wondering what exactly it is about these young people that gives you hope. Do you feel like this generation of young people is different from other generations?"

"In terms of the environment and social justice," Jane said, "this generation *is* different. When I was growing up there was nothing taught about these issues in school. But gradually more activists began to write about them. One of the most significant books that influenced people back in the 1960s was Rachel Carson's *Silent Spring*, about the horrific damage caused by the use of DDT."

"That book really did help start a movement," I agreed. "The right book or the right film at the right time really can change the culture. Al Gore's *An Inconvenient Truth* is another example. Books like Michelle Alexander's *The New Jim Crow* and Bryan Stevenson's *Just Mercy* have helped create a criminal justice reform movement in the United States."

"Yes, it's true. Gradually, over the past sixty years, these issues have been brought out into the open—and some schools began to include awareness of environmental and social issues in the curriculum. And nowadays even if they're not taught at school, they're in the news, on television—everywhere. Children can't escape from hearing about the climate crisis—pollution, deforestation, biodiversity loss—and increasingly about our social crises—racism, inequality, poverty. So young people are now much better equipped than we were to understand and deal with the problems we've created. And to understand how all these issues are connected."

"Indeed," I said, "it's great that we are educating future generations to be more environmentally and socially aware and that they are changing their parents, but we have massive challenges *now*. We need change makers in power right now. We don't have the time to wait for those young people to grow up—"

"A lot of them have already grown up," Jane countered. "There are three decades of Roots & Shoots alumni now, who have taken the values they acquired as members into their adult lives."

Still, I wasn't convinced. "I hear that, but I think that a lot of people push back on saying that youth are the solution. After all, most past members of Roots & Shoots are not in positions of power yet. We need the president of the United States—who's not going to be a twenty-year-old or thirty-year-old—to lead the way. We need all of the people to be dealing with this in the next ten years."

Jane didn't miss a beat. "That's true. But it's going to be the twenty-year-olds and thirty-year-olds who will vote in the right president."

Once again, Jane was prescient. Eleven months later, an increase in young voter turnout would help vote out Donald Trump, who had pulled the United States out of the Paris Agreement, and elect Joe Biden—and one of the first major acts of his presidency would be to rejoin the Paris Agreement and recommit to building a healthier economy and planet. Sixty-one percent of Americans age eighteen to twenty-nine—who made up almost one-fifth of the electorate— would vote for Biden. Although Biden would earn over seven million more votes than his opponent, in the bizarre math of the Electoral College, the election would ultimately be decided by just a few hundred thousand votes in key battleground states. It would be the votes of the Jane Generation that would guide the globe's largest superpower in the right direction. But that was all in the future as we talked that day in the cabin in the woods, and at the time, all I could say to Jane was, "Let's hope you are right."

Jane leaned forward and stirred the dimming embers of our fire, and we watched the flames brighten again.

"And there's another thing," Jane said, as she settled back in her chair. "A number of those Roots & Shoots alumni I was talking about have gone into politics. And others are businesspeople, journalists, teachers, gardeners, urban planners, parents—you name it. Many are now working for the environment in some capacity—including the minister of the environment in the Democratic Republic of the

Three members in Tanzania. This T-shirt epitomizes the Roots & Shoots values. (JANE GOODALL INSTITUTE/CHASE PICKERING)

Congo, who had been in a Roots & Shoots group at school. He is really trying to curb the illegal bushmeat trade and animal trafficking in his country."

Jane said that young people today are not just better informed but they're becoming more engaged in decision-making and in the political process. Roots & Shoots, for example, is more than an environmental program. It's actually teaching people the values of participation and democracy. Joint discussions, joint decisions, doing things together.

"The full impact of youth empowerment programs in these countries hasn't become apparent," she said. "Yet."

Jane's "yet" was a powerful reminder that even the most hopeless circumstances can, in time, change.

It also reminded me of Stanford professor Carol Dweck's use of the words "not yet" to identify a growth mindset, or the belief that we can change and grow. Children—and adults—who have a growth mindset are much more successful than those who have a fixed mindset about themselves and the world. But could small educational programs really stand up to the might of totalitarian regimes and vested business interests?

"In many countries," I said, "you can't fight the government or speak out against injustice for fear you'll get put in prison or killed. What do you say to young people in these countries?"

"I tell them that while they will have to live with the existing system, they can still hang on to their values, make some small difference every day, and maintain their hopes for a better future."

It was almost as if Jane was saying that our collective hopes and dreams, even when they cannot be realized, have power, perhaps waiting for the right time to be realized. Even so, my New York skepticism was triggered. "That's wonderful, but doesn't it feel like a drop in the ocean, given the overwhelming autocracy or tyranny that people are facing around the world?"

"But millions of drops actually make the ocean."

I smiled. Hope, checkmate.

Nurturing the Future

It was getting toward evening, the sun sinking fast, as I continued to think about all the major problems that had been denied or ignored for so many years. Of all the people who denied climate change, of cultures that taught children that boys were superior to girls, of all the misguided adults who taught children that some races or groups were better than others. How fear, prejudice, and hatred can be taught as easily as courage, equality, and love. "So how do we change these entrenched worldviews fast enough?" I asked.

"Oh, Doug, I honestly don't know. My hope is that there are more and more people who are concerned, more and more programs working on these issues. Trying to alleviate poverty, improve social justice, fight for human and animal rights. And more and more children are becoming involved when they are very young."

She paused to reflect, and soon her eyes lit up with another hopeful story.

"I'm thinking of Genesis," she said, "a young American girl, whose favorite food, when she was six years old, was chicken nuggets. One day she asked where they came from. Her mother tried to fob her off, telling her they came from the store. 'But where does the store get them?' So her mother told her and not only did Genesis stop eating her favorite food, but she found out as much as she could and now, aged thirteen, she gives talks about the importance—for animals, the environment, and human health—of becoming vegan. There are so many examples of very young children becoming activists. And those who are most committed and successful usually have supportive parents."

Once again, I thought about my own children and wondered how my own actions affected their worldviews. "How do we as parents nurture our children to be hopeful and to be ready for the future they will encounter?"

"To begin with, I learned from the chimps the importance of the first couple of years of life," Jane replied. "After sixty years of research it is very clear that the young chimps who had supportive mothers have tended to be the most successful. The males got higher in the dominance hierarchy—they were more self-confident and tended to sire more offspring—and the females were better mothers."

"So how would we translate that into human parenting?" I asked Jane.

"Well, it's not that different. I also did quite a bit of research on raising human infants when I was working on my Ph.D. thesis. And it is obvious that what is important for our children is that during the first couple of years they receive care and love from at least one person who is always there for them. They need reliable and supportive caregiving. And it does not have to be the biological mother or father, or even a family member."

"So many parents think that *supportive* parenting means permissive parenting," I said. "Where does discipline come in?"

"Discipline is important, but I believe it is crucial that a small child is not punished for something that he or she has not been gently taught is wrong," Jane said. "I saw a mother beat her two-year-old for spilling a bit of the milk he did not want and making patterns on his tray with his finger. Yet his behavior was merely a demonstration of how children learn about the world around them, the properties of things. He did not deserve the harsh punishment he received. Physical punishment is wrong. Chimpanzee mothers distract their small infants from undesired behavior by tickling or grooming."

I loved this image of chimpanzee mothers tickling or grooming their children and thought about how I had often tried to change a mood to change a mind when one of my three young children was melting down.

"What can we do about the young people who grow up without support, perhaps in an abusive home?"

As usual, Jane answered with a story.

"I had a letter from a fourteen-year-old the other day who was in a juvenile detention center. She wrote, 'My life was a mess and I was on drugs, and I came here and I hated it. And then in the library I found a copy of *My Life with the Chimpanzees*. I never had a supportive mother, but when I read that book, I thought Jane can be my mother.'

"Her mother had never told her she could succeed. But when she read how my mother had supported me, and the difference that had made, she started to realize that she, too, could follow her dreams. I would be her role model—that's what she meant by saying I could be her mother. She started behaving well, working hard—she turned her life around."

I thought about this young woman, about the power of books and stories and role models to change a child's life. And I thought about what Jane had said about how important our environment is and that our human nature is adaptable enough to fit into the world in which we must survive. How we can nurture our children is so very dependent on the larger community in which we live. There can be little doubt that the poverty, addiction, and hopelessness surrounding Robert White Mountain's son contributed to his dying by suicide at sixteen.

I told Jane about a hope researcher named Chan Hellman who grew up in poverty in rural Oklahoma. His father was a drug dealer and would take Chan along on his deals to reduce the chance of violence. By the time Chan was in seventh grade, his father had moved away and his mother, who'd been hospitalized several times for depression, stopped coming home. Chan was eating only one meal a day—the lunch served at school—and living alone in a house where the electricity had been cut off.

"One night, he was in that dark house feeling such despair and hopelessness that he got his parents' gun and placed the barrel under his chin. Then he flashed on a memory of his science teacher, who was also his basketball coach, telling him, 'You are going to be all right, Chan.' He thought about his teacher's words and about how this man clearly cared about him and believed in him. It was then he decided that maybe his future could be better and put away the gun."

"Do you know what happened to Chan?" Jane asked.

"He is now a man in his fifties with a loving wife and family and a successful career as a hope researcher focusing on abused and neglected children. A couple of years ago he met up with his old teacher. He told the teacher how he had saved his, Chan's, life. The teacher had no recollection of his lifesaving words. Chan says it is a reminder of how much our words matter, even when we don't know it, and that the real takeaway is that hope is a social gift."

From talking with Jane and doing my own research, I was starting to see that hope is an innate survival trait that seems to exist in every child's head and heart; but even so, it needs to be encouraged and cultivated. If it is, hope can take root, even in the grimmest of situations, one of which Jane had witnessed firsthand.

"I want to tell you about a Roots & Shoots group that began in Burundi," Jane said. "Burundi is just south of Rwanda, and the genocide of the Hutus also happened there. As we already discussed, the Rwandan recovery from genocide is a great source of hope, but it came about because of the international aid that poured in after the visit by President Bill Clinton."

"I remember the horror of the genocide and the extraordinary efforts in Rwanda to forgive and heal," I said.

"But as I also mentioned, Burundi got nothing, absolutely nothing. It was more or less ignored by the international community and left to work things out for itself. Not surprisingly, it has not recovered in the same way and there is intermittent unrest and violence. The first Roots & Shoots group there was started by a young Congolese man whose whole family had been massacred and who had escaped over the lake to Kigoma in Tanzania. At the school he went to, there was a Roots & Shoots group; and several years later, when he was visiting Burundi, he decided to start a Roots & Shoots program there. It began with four ex–child soldiers and five women who'd been raped. I remember sitting around the table with them and listening to them talk about what they had endured.

"None of them went into details—instead they seemed withdrawn, but I could see the pain in their eyes. As I have done so often, I tried to imagine being treated like those young women and the countless other women who have experienced unthinkable abuse. Some, of course, never do recover. But despite all they had suffered, these young Burundians wanted to help others recover from their own traumas and show them that there was a way forward. I was amazed by this example of the indomitable human spirit that we find in young people everywhere."

Jane told me that the program had spread throughout Burundi, and not long after that evening we spent by the fireplace, she sent me a recent batch of letters she had received from Roots & Shoots members there. One child, Juslaine, wrote, "A long time ago Burundian people didn't know the importance of working together, but now we are working together as one family due to the seminars given by the leaders of Roots & Shoots Burundi." Another boy, Oscar, wrote, "We no more live in troubles because each year we celebrate the international peace day. Now we are living peacefully with our neighbors."

Jane told me that one of the ex–child soldiers, David Ninteretse, inspired many volunteers from the communities to initiate Tacare-type programs that encouraged people to start small businesses. He also got volunteers to start Roots & Shoots groups in schools, many of which decided to plant trees to combat deforestation. A boy named Eduard said, "My village was like a desert, but now trees are found everywhere, and the rain comes regularly." Other children commented on how there are no more forest fires, how the air is fresh, and the animals have returned to the forest, as the hunting stopped.

"You see," said Jane, "they learn that everything is interconnected, and that their community is not just the people around them but also the animals and plants and the land itself."

I thought of what Robert White Mountain had said about how his tribe was at one time the caretakers of the land but that over the years they had lost that connection. Jane said she had heard that he was trying to revive that connection by creating a large community garden. I thought about the young girl in juvenile detention, who had turned her life around after reading Jane's book, and about Hellman, who had survived shocking neglect. I thought about how important it is to nurture young people in a way that cultivates hope and empowers them to meet the challenges of the future. They are certainly inheriting many. While I was convinced that young people were an important reason for hope, I could see clearly our adult responsibility to leave them the most thriving and sustainable world.

It was late evening now and we still had one more reason for hope to cover. Jane suggested we pause our conversation here and pick it up again in the morning. It was hard for me to stop. I'd been looking forward to discussing Jane's next reason for hope—a reason

Children in the Republic of the Congo on UN International Day of Peace, flying a Giant Peace Dove (Roots & Shoots groups do this all over the world, using old bedsheets) and going to a tree planting project. (JANE GOODALL INSTITUTE/FERNANDO TURMO)

we can still find even when there seemingly is none. I agreed to return in the morning and set off in the dark night to my lodge where I was staying near Jane's cabin.

REASON 4: THE INDOMITABLE HUMAN SPIRIT

I met up with Jane the next morning. Jane was staying in the cabin with Patrick van Veen, the Jane Goodall Institute's global chairman, and his wife, Daniëlle, and their two dogs. I joined her in waving goodbye to Patrick and Daniëlle, who again had kindly agreed to go off for the day with the dogs to give Jane and me time alone. Once again, we settled down by the fire with our mugs of coffee, eager to resume our conversation.

"I was thinking about this last night," I began, "and I thought that before we talk about your last reason for hope—the indomitable human spirit—I'm interested to know how you define 'spirit.'"

Jane thought for a moment and then she said, "Nobody's ever asked me that before. I think different people will define this very differently depending on their upbringing, education, and religion. I can only tell you what it means to me. It's my energy force, an inner strength that comes from my sense that I am connected to the great spiritual power that I feel so strongly—especially when I'm in nature."

I asked Jane if that sense of "the great spiritual power" comes to her especially when she is in Gombe.

She nodded. "Oh yes, absolutely. And once when I was alone in the forest, I suddenly thought that perhaps there was a spark of that spiritual power in all life. We humans, with our passion for defining things, have named that spark in ourselves as our soul or spirit or psyche. But as I sat there, embraced by all the wonder of the forest, it seemed that that spark animated everything from the butterflies that fluttered past to the giant trees with their garlands of vines.

"When we were discussing the human intellect the other day, I spoke about how indigenous people, including many Native Americans, talk about the Creator and see animals and flowers and trees and even rocks as their brothers and sisters. I love this way of seeing life."

I was intrigued by that description and wondered aloud how the world might be different if all humans saw other beings and even rocks as worthy of respect and care as our siblings would be.

"A better world, I imagine," Jane said. "But of course we don't really know how it would be different. At least not yet."

I couldn't help but smile at Jane's reliably hopeful "yet" at the end of the sentence, which brought me back to the discussion of the day before. "What do you mean by the *indomitable* human spirit?" I asked. "And why does it give you hope?"

Jane stared into the fire for a few seconds before answering. "It's that quality in us that makes us tackle what seems impossible and never give up. Despite the odds, despite the scorn or mocking of others, despite possible failure. The grit and determination to overcome personal problems, physical disability, abuse, discrimination. The inner strength and courage to pursue a goal at any cost to self in a fight for justice and for freedom. Even when it means paying the ultimate price of giving up one's life."

"Do you have favorite examples of people who embody this spirit?" I asked.

"Some people immediately spring to mind. Martin Luther King Jr., who fought for an end to discrimination and income inequality and preached nonviolence, despite terrible adversity. Nelson Mandela, who was imprisoned for twenty-seven years for his fight for an end to apartheid in South Africa. Ken Saro-Wiwa, a Nigerian who led nonviolent demonstrations against the pollution of the land by Royal Dutch Shell and was executed by his government." I could not help thinking of the story of the CEO of Royal Dutch Shell and his transformation, as well as the dark history of so many of the oil and gas companies that had endangered the planet.

Jane continued, "Winston Churchill, of course, who inspired Britain to fight Nazi Germany even when almost every European country had been defeated. Mahatma Gandhi, the Indian lawyer who led the nonviolent movement that finally ended British colonial rule. And the example that will spring to the minds of Christians, of course, is Jesus. I have been so deeply inspired by these people who demonstrate, in their lives, this indomitable spirit. The influence they have had on the course of history—well, I could not begin to evaluate it. And they are just a few examples."

"So," I asked, "the indomitable human spirit is what helps us to go on even when it seems hopeless. And something that inspires others?"

"Yes, exactly. And in addition to those icons who have inspired millions, there are all those among us who face really daunting social or physical problems in their own lives. The refugees who endure great danger and hardship to escape from violence and, knowing no one, manage to create a life for themselves—even when, as is sadly so often the case, they face discrimination on arriving, finally, at their destination. People with disabilities who refuse to let this prevent them from following their dreams. And they, too, inspire all those who come in contact with them by their courage and determination to overcome their challenges."

When I Decide to Climb Everest

"Do you think it's our indomitable spirit that has allowed us to survive and thrive?" I asked. "After all, we humans are the physically weakest of the apes."

"Well, no, it was our brains and our ability to cooperate, along with our adaptability, that's enabled us to be successful," Jane said. "Our indomitable spirit has taken us a bit farther, I suppose. Because we're in the unique position of being able to understand exactly what may result from a conscious decision to tackle what we are told is an utterly impossible course of action."

"Do you believe the chimps have an indomitable chimp spirit?"

Jane chuckled. "They certainly have the will to live, as described by that great humanitarian doctor Albert Schweitzer. The will that causes them to struggle to recover from sickness and injury and other challenges like so many animals—so long as they are psychologically healthy. Animals, like us, can feel helpless and hopeless; and in this state of despair they may give up when challenged by illness, injury, or some traumatic event—like being captured, for example.

Some infant chimps survive horrific situations, while others may give up and die even when their situation is far less awful."

"But you think this will to live is different from the indomitable spirit that you're describing for humans?" I asked.

"Well, I think for us it's more than the will to live when confronted by a life-threatening situation—although we certainly share that with other animals. It's an ability to deliberately tackle what may seem to be an impossible task. And not give up even though we know there is a chance we may not succeed. Even when we know it may lead to our death."

"So this indomitable spirit requires the amazing human intellect and imagination—and, of course, hope?"

"Yes," said Jane, "and it also requires determination and resilience and courage."

I told Jane I had a very important role model in my life who embodied that indomitable spirit: my grandfather. "He lost his leg as a boy," I said. "And even with a wooden leg he became a ballroom dancer and competitive tennis player! He became a neurosurgeon and performed a pioneering separation of conjoined twins, which he had been told was impossible. During World War Two, he would show the recent amputees how to live with a prosthetic and assure them that they could have a full life. He had a motto: 'The difficult is hard, the impossible just a little harder.'"

"That's a wonderful example of a human's indomitable spirit," Jane said. "That's exactly it."

"Derek is another example," I said, referring to her late husband.

"Yes, Derek is another marvelous example of resilience, grit, and an indomitable spirit," she said. "He was in the Royal Air Force fly-

Doug's grandfather, Hippolyte Marcus Wertheim, leaving York Hospital on December 7, 1936, after performing a successful separation of conjoined twins, a surgery he had been told was impossible. He walked with a limp because of his prosthetic leg.

ing a Hurricane and was shot down over Egypt when the Allies were fighting Rommel, the Desert Fox, in World War Two. He survived the crash and was rescued, but his legs were partially paralyzed because of the damage to the nervous system at the base of his spine where he had been hit by a German bullet.

"Derek's doctors told him that he would never walk again, but

Derek was badly damaged in World War Two when his plane was shot down. He was told he would never walk again. He was determined to prove the doctors wrong and he succeeded! (JANE GOODALL INSTITUTE/ COURTESY OF THE GOODALL FAMILY)

he was determined to prove them wrong and he never gave up. That he was eventually able to walk with the use of only a single cane was nothing short of a miracle. One leg was almost entirely paralyzed—he had to sort of swing it forward with his hand for each step. And Olly, my aunt, who was a physiotherapist, examined him and said, 'Well, actually, anatomically, looking at all the muscles and everything, he shouldn't really be able to use the other leg either. He's walking by sheer willpower.'"

"That's so inspiring," I said. "It makes me think of my father's accident, too. He fell down a flight of stairs, almost exactly five years before he died. My dad suffered a very serious traumatic brain injury that left him delirious for over a month. We were told that he might never come back or be himself again. When he finally came back to his right mind, my brother said he was sorry my dad had endured such a traumatic experience. My dad responded, 'Oh, no, not at all. It's all part of my curriculum.'"

"What a great way to put it," Jane said. "Yes—all life's challenges are like our own individual curriculums that we must work hard to follow and master."

"In that small perceptual shift, my dad had been able to reframe a negative experience more positively and to find meaning in it," I said. "The fall and recovery were harrowing, and yet the last five years of his life were filled with deep psychological growth and even richer relationships with his family and friends. Archbishop Tutu once explained to me that suffering can either embitter us or ennoble us, and it tends to ennoble us if we are able to make meaning out of our suffering and use it for the benefit of others."

"Yes." Jane nodded. "And I know your son has recently been in a serious accident and has also been quite resilient," Jane said, her voice etched with concern.

It was true. My son, Jesse, had been in a surfing accident a month before Jane and I met in Tanzania and was suffering from his own traumatic brain injury as well as a broken occipital bone. "He has been in extreme pain but has also shown remarkable resiliency and hopefulness. And, as the research on resilience shows, having a sense of humor helps. Jesse actually started doing stand-up comedy as part of his healing."

"Yes, having a sense of humor really is helpful," Jane said. "I remember a story Derek told me. He was just out of hospital and on crutches. He had to meet someone at the Ritz Hotel. When he sat, he forgot both his legs were in plaster casts—they shot out stiffly in front

Chris Koch, a personal hero, and a perfect example of an indomitable spirit. (JANE GOODALL INSTITUTE/SUSANA NAME)

of him"—Jane demonstrated for me, kicking both her legs straight out—"knocking over the table so that teapot, cups, milk—the lot—flew in all directions. There was a moment of shocked embarrassment, then Derek began to laugh and soon the whole table, and even the dignified waiter and people from nearby tables, joined in."

I thought about all those other people that I had heard about who had overcome personal disaster, all those whose lives are inspirational, who illustrate the indomitable human spirit. I asked Jane if she had any more examples to share with me.

Jane told me about Chris Koch, a Canadian who was born with no arms and no legs—just short stumps for arms and one extremely short leg stump. He gets around on a long board—and there is virtually nothing he can't do. He travels on his own around the world, goes in for marathons, drives tractors—and is an excellent inspirational speaker.

"His parents never told him he couldn't do what his brothers and sisters did," Jane explained. "They always told him he could do anything. They never said, 'Oh, you can't do that.' His eyes shine with intelligence and a love of life. I asked him whether people have offered him prosthetic limbs and he replied, 'Well, yes they have, but I think that I've been put together this way for a purpose. I think I'll stay as I am.' But then after a pause, and with a twinkle in his eyes, he said, 'But I might take them up on it when I decide to climb Mount Everest.'"

As we sipped our coffee and told these stories, I felt uplifted just by talking about these examples of the indomitable human spirit and thinking about their hopefulness and bravery.

The Spirit That Never Surrenders

"Earlier you said Churchill provided an example of an indomitable human spirit," I said. "Can you say more about how he influenced you and others during World War Two?"

"Yes, indeed I can," Jane answered. "It was Churchill's indomitable spirit and his belief in the British people that inspired them and called out their courage and their determination not to be beaten by Hitler.

"I think the whole experience of growing up during that war helped to shape who I am," she continued. "Even though I was only five years old when it began, I knew or sensed what was going on. I felt the atmosphere. Everything seemed bleak and hopeless—after all, for a time Britain stood alone after the occupation or surrender of most of the other European countries. Our army wasn't prepared.

Our navy wasn't prepared. Our air force was puny, compared with the Luftwaffe."

I remembered reading about this terrible time in history when it looked like Hitler would win the war and occupy Britain; and, listening to Jane, who'd lived through this moment, I could sense the fear the people of Britain must have felt.

"Cutting through the despair came Churchill's speeches," Jane continued, "sharing his belief that Britain would never be defeated, and the speeches brought forth the fighting spirit in the British people. His most famous speech was delivered when Germany had defeated and overrun much of Europe and things looked really bad for the Allied forces. But Churchill inspired people with his rousing words, saying that we would defend our island to the very end, we would never give up, we would fight the enemy on the beaches and in the fields and in the hills and in the streets. We would never surrender. There was tumultuous applause at the end of this speech; and during this someone overheard Churchill mutter to a friend, 'And we'll fight them with the butt ends of broken beer bottles because that's bloody well all we've got.'"

Jane chuckled. "He had a great sense of humor, a very British sense of humor.

"And he didn't hide away from what was going on. During the terrible weeks of the Blitz, when London was bombed night after night, he often went to give words of encouragement to the people sheltering in the Underground stations, people who were shocked by the loss of life, the screams of the injured, the destruction of their homes. He inspired a renewed determination in everyone to go on fighting Hitler to the bitter end."

Jane told me her memory of the Battle of Britain, about all those young British pilots, joined by those from Canada, Australia, and

Poland, in their Spitfires and Hurricanes, who risked their lives day after day; when so many died as they fought the superior strength of the Luftwaffe. It was a decisive moment in the war. Hitler had realized Germany could not gain mastery over the sea until his Luftwaffe destroyed the British air force. And when it became clear this was not happening, and that he could not defeat the Royal Air Force or the morale of the British people, he called the air attack off.

"Churchill's famous words about the RAF still bring tears to my eyes," said Jane. "After all the acts of heroism and the tragic loss of young lives—'Never in the field of human conflict was so much owed by so many to so few.'

"So many people died in that war, Doug. Not only in the armed forces but thousands of civilians caught up in the fighting and the bombing. And not just the Allies—the German people, too."

We were both silent for a moment, absorbing the truth of what Jane had just said and honoring the many who'd died. "Looking back now, what do you think the lasting lesson was that you took from the war once it was over?" I asked.

"Well, it goes right back to what we're discussing," Jane said. "I was beginning to understand what people are capable of and how an indomitable determination can motivate and inspire a nation and turn what seems an inevitable defeat into victory; that with courage and determination the impossible becomes possible."

I paused the recording there and we decided it was time for a quick stretch and a little more coffee. As I refilled our mugs, I noticed Jane was studying the floor where a shaft of morning sun was lighting up the pattern on the carpet. "What are you thinking about?" I asked, starting the recording again.

"I was just thinking about how disaster and danger can bring out the best in people. World War Two created so many heroes, those who risked their lives to save their comrades or their battalions—all those Victoria Crosses for bravery. So many of them awarded posthumously. The resistance fighters, men and women, who went undercover to fight the Nazis in any way they could—and, by the way, many of them were Germans. And when they were discovered, they so often refused to give away the names of others in the network, even when they were tortured. I used to lie awake, feeling sure I would not have the courage to keep silent while my fingernails were torn out, praying I would never be put to the test. And there were all those who risked their lives to help the Jews escape or hid them in their homes. And the quiet heroism of the citizens of London who endured the Blitz and helped each other. They displayed their grit and Cockney sense of humor day after day as their houses were destroyed around them."

"I think this is always the case—disasters always lead to stories of altruism and bravery," I said. "I'll never forget seeing the firefighters running into the flaming, collapsing buildings on 9/11 as the terrified, dust-covered people were running out. Or seeing all the international relief workers rushing to help when there's an earthquake or devastating hurricane. And during last summer, witnessing all the people battling the wildfires and rescuing the trapped humans and animals in Australia and then California."

"Yes," said Jane, "all these stories of heroism, of courage, of self-sacrifice illustrate the indomitable spirit that is so often revealed by danger. Of course, it is there all the time, but so often nothing happens to call upon it."

"I suppose examples of the indomitable human spirit rallying us to 'fight the unbeatable foe' and 'right the unrightable wrong' have been with us throughout history."

"Indeed," Jane said, "you only have to think of David and Goliath. And another picture comes to mind of that lone man confronting the tanks of the Chinese Army in Tiananmen Square with his shopping bags. Both of those examples seem to symbolize the indomitable courage that is sometimes shown when people stand up against a seemingly invincible force.

"And there are all the indigenous people in so many parts of South America who are standing up against the vested interests of governments and big business as they try to defend their traditional lands from logging and mining. They are prepared to sacrifice their lives, and they often do."

"It's true," I said. "Even as we've witnessed horrifying acts of cruelty and self-interest in politics lately, there have always been people who are willing to risk imprisonment, beating, torture, and even death in order to resist tyranny, injustice, or prejudice."

"Yes," Jane said, "think of the early suffragette movement in England, led by Mrs. Emmeline Pankhurst, when the women tied themselves to the railings outside the House of Commons as they fought for women's right to vote. And think of the number of people, all over the world, who have tied themselves to trees, or climbed up into the branches, to try to protect a forest from the bulldozers."

"Another inspiring example is Standing Rock," I said, referencing the 2016 protests to stop construction of the Dakota Access Pipeline that would likely threaten the Standing Rock Sioux Reservation's primary water source and desecrate their sacred sites. "The police used pepper spray, tear gas, rubber bullets, and even sprayed the protestors with water in the freezing winter, and still the protestors stayed. Thinking about that now, it was the young people of Standing Rock who emerged as leaders in that occupation."

"Oh, Doug, there are so many unsung heroes," Jane said. "So many examples of the indomitable spirit, the spirit that will never give up or give in, and so many examples that will probably never be told. There are the pacifists, who risk ridicule by refusing to fight for their country but risk their lives daily by driving ambulances into the thick of battle to rescue the wounded. The journalists who risk their liberty and their lives to speak the truth about corruption and brutality in tyrannical regimes, whistleblowers who feel compelled to reveal the truth of abominations that go on behind the closed doors of powerful corporations, courageous people who secretly film what goes on inside the factory animal farms or capture scenes of brutality on the streets.

"And I love the story of Rick Swope, who risked his life when he rescued a chimpanzee, Jo-Jo, from drowning in the moat that surrounded his zoo enclosure. Jo-Jo, an adult male who had lived alone for many years, was being introduced to a large group. When one of the high-ranking males charged toward him, asserting his dominance, Jo-Jo was so terrified he managed to scramble over the barrier erected to prevent the chimps from drowning in the deep water of the moat that surrounded the enclosure.

"As you probably know, chimps can't swim. Jo-Jo disappeared under the water, came up gasping for breath, then disappeared. There were several people, including a keeper, watching—but only Rick jumped in, as his horrified wife and three children looked on! He managed to grab hold of the big male, somehow carried him over the barrier, and pushed him up onto the bank of the enclosure. By this time three of the big male chimps were charging down, hair bristling, and Rick turned to get back over the railing. Jo-Jo was alive but weak, and he started to slide back into the water. In the video shot by a visitor you see Rick pause. He looked at his wife and children and the keeper, who

Image from video taken of Rick Swope rescuing Jo-Jo
after he fell in the zoo's moat. (YOUTUBE)

were all yelling at him to leave the moat. Then he looked to where
Jo-Jo was disappearing under the water. And he went back, pushed him
back up again, and stayed until the chimp had grabbed a tuft of grass
and managed to pull himself onto level ground. Fortunately, the three
male chimps just watched.

"Later Rick was interviewed. 'You must have known it was
dangerous—why did you do it?' he was asked. 'Well, you see, I happened to look into his eyes, and it was like looking into the eyes of a
man,' he said. 'And the message was, "Won't anybody help me?"' That
same look in the eyes of the vulnerable and oppressed people that has
appealed to human altruism led to so many heroic acts."

"That is an incredible story," I said. "Clearly Rick's actions prove
that our moral code extends way, way beyond helping one's kin, and
he could hardly expect Jo-Jo to reciprocate! I think the story so well
illustrates the courage and the respect for life that is needed to make
change in our society. Do you think that this kind of respect and
courage can help us overcome the many problems in our society?"

"I'm absolutely sure that it can help," Jane answered. "Of course, there is one problem—the same courage and selflessness can be shown by people who have been brainwashed—think of the suicide bombers who believe they will be rewarded in paradise for blowing up innocent people. In fact, there are heroic actions performed by people on both sides of an issue. It points to the importance of the cultural and religious environments in which people are raised."

"But when it comes to the grim environmental situation we face today," I said, "do you think we could all come together and use this same energy and determination to tackle climate change and loss of biodiversity?"

Jane didn't answer immediately—she was obviously gathering her thoughts. "There is no doubt in my mind that we could. The trouble is that not enough people realize the magnitude of the danger that we are facing—a danger that threatens to utterly destroy our world. How do we get people to heed the dire warnings of the people on the ground who have been fighting this danger for so long? How do we get them to take action?"

Jane was looking deeply concerned.

"That is why I travel around the world—trying to wake people up, make them aware of the danger, yet at the same time assure everyone there is a window of time when our actions can start healing the harm that we have inflicted. Using our brains, counting on the resilience of nature. Urging everyone to take action by first describing the very real danger, but then emphasizing that we still have a window of time, that there really is reason to hope that we can succeed."

"We've talked a lot about the resilience of nature, and this makes me curious whether the indomitable human spirit is linked to resilience."

"Well, of course—after all, everything is interrelated," Jane said. "So while the courage of the indomitable spirit is often revealed in times of disaster, as we have said, this is not true for everyone. Some people go under. And I do think this is linked to resilience, and if we are optimistic or pessimistic."

Nurturing the Indomitable Spirit in Children

The winter sun continued to illuminate the cabin. As Jane thought about the connection between resilience and the indomitable human spirit, I was wondering if children could be taught—or at least helped—to become more indomitable so that they would be better able to cope with life's inevitable challenges as they grew up. The parents of Chris Koch, the man who was born without arms or legs, had done this brilliantly. They gave him the self-confidence and mental strength to succeed. I mentioned Chris's example to Jane.

"Oh yes, I'm sure that self-confidence is part of resilience and one's upbringing plays such an important role," she said. "When I think of the other children who have overcome physical disabilities, they almost always had support from one or both parents or some other adult who was 'there for them.'"

"And, of course, while some people face physical adversity, like Chris and Derek, and like my father and son," I said, "there are some who fight and overcome trauma induced by war or childhood abuse or domestic violence—trauma that leaves psychological scars."

"I suppose that in all these cases there will be resilient people who overcome both physical and psychological trauma, whilst others simply don't have that resilience. It's not always clear as to why. Perhaps some people who are genetically predisposed to pessimism don't have a loving enough upbringing to foster resilience and hopefulness."

I shared with Jane how the research on resilience has interesting resonance with the research on hope. Psychological resilience is the ability to cope with crises and to remain calm and move on from such incidents without long-term negative consequences. Like a resilient ecosystem that recovers after a natural disaster or human-made disturbance, resilient people are able to recover—though it may take time depending on the severity of the trauma.

"Overall, a resilient person is able to bounce back—or even bounce forward—stronger as a result of the adversity," I said. "These people are more hopeful, and they may see challenges as opportunities."

"It's really sad," Jane said, "how some people manage to cope, and cope in the most amazing way, while others give up, become bitter and depressed—may even kill or seek to kill themselves, especially if they don't have family or friends to help them."

"There may be some exceptions," I said, "but overall I think we can agree how important consistent nurturing, safety, and care is when it comes to fostering resiliency in children. From what you've seen, do you think that's true for chimps as well?"

"I do," Jane replied. "We've known chimpanzees who've been taken from their mothers and abused as infants—some who were trained using severe punishment to perform in entertainment, some who were confined in bare cages in medical research labs—who have never really recovered after being rescued, never been able to fit into a normal chimpanzee group. And they may show what is surely post-traumatic stress disorder. There was one female, who every so often would stare off into the distance and scream and scream hysterically. She had been separated from her mother as an infant and raised in a lab setting where she had been deprived of love. By contrast, when traumatized infants whose mothers were shot in the wild arrive at one of our sanctuaries and are *immediately* given love and care, they usually bounce back fairly quickly."

How the Indomitable Human Spirit Helps Us Heal

"It's wonderful to know how universal this resilience can be," I said. "I'm also struck by the examples you shared yesterday—how people who have been horribly abused can sometimes overcome their trauma and then devote themselves to helping others who are still struggling."

"Yes," said Jane. "You mean the young women in Burundi who had been captured and raped, and the young men who had been forced to become child soldiers. Through counseling they were able to face what had happened, find the strength to move forward with their lives, and then decide they wanted to use their experience to help others who are having trouble finding their way out of despair or anger. And, of course, it helps your own healing if you are doing something to help others."

Jane said she gets "a good many letters" from people who are trying to cope with adversity—sometimes it's from the parents of children with life-threatening or incurable diseases, or from people who were abused as children and are still trying to cope, and often from people who have lost hope because of the damage to the environment. She told me she often has phone calls with or is writing to people with physical or mental problems.

"And what are they wanting from you?" I asked.

"They want help, support," she answered. "It is a huge responsibility, and, to be honest, it's sometimes draining. At the same time, it is a privilege, because so often they say it really does help them when they talk to me. Even to hear my voice, which they say is calming and brings peace. I don't understand why this is so but have come to accept it as a gift I have been given. And I feel compelled to use this gift. It has given me a real understanding of the kinds of hardships and traumas that people have to face, and a real admiration for the way that people cope with what has happened with determination and courage. It's that indomitable spirit again!"

Jane told me about a young woman who wrote her a letter that enclosed a police notice asking for information that would help them locate a missing person.

"I'll call her Anne," said Jane. "The missing woman was Anne's adored older sister, who had last been seen as a teenager getting into a car with a man in a gas station during a terrible storm. That was thirty-two years ago."

Jane said Anne worshiped her older sister, who had been one of the few stabilizing influences during her troubled childhood.

"When I met Anne," Jane continued, "she was not very coherent, but in the letter that she handed me, and I later read, she asked if I could sign a petition for getting her sister's case reopened. Her writing was so tiny that I almost needed a magnifying glass. I wrote back to her, and she told me she'd given a similar letter to maybe forty people. 'But you're the only one who's written back to me,' she said."

They began a correspondence and eventually Jane gave Anne her phone number.

"She would call me three or four times in a row, and she was always crying at the start of a conversation. Each time her voice was quite different. I'd read a good deal about the strange disorders of the mind, and I realized that she had developed multiple personalities, which is a recognized way of dealing with extreme trauma."

Jane went on to explain the horrific trauma Anne had experienced. When she was two years old, her father returned from the Vietnam War and began physically abusing his wife, who became chronically depressed and had to be put into treatment. Anne and her sister then went to live with their father, who had remarried. For the next ten years or so, Anne was horrendously sexually abused by her father and physically abused by both him and his new wife. Her sister, for some reason, escaped this treatment. When, eventually, Anne's mother was discharged, she made a home for Anne, who was twelve years old, and her sister. And then, just after Anne got a taste of normal family life, came the horrible disappearance of her sister when she was on her way home to celebrate Thanksgiving. No wonder Anne was in such a terrible state.

"It was quite incredible," Jane told me. "She had *twenty-two* distinct identities—as she came to trust me, she actually wrote out the three family trees of the various personalities who ranged from small children to adults. And, as I said, when she called me—which was very often—Anne spoke in different voices. Sometimes she would hang up—then call back speaking in a quite different voice—perhaps the voice of a small child. I'd ask her, 'So who are you this time,

Anne?' Finally, I encouraged her to write the details of the horrible abuse she had suffered."

Then, concerned that she'd offered advice she wasn't qualified to give, Jane wrote to Dr. Oliver Sacks, the eminent neurologist who specialized in disorders of the mind.

"I explained Anne's strange case to him and that I'd told her she should write down some of her terrible experiences. 'But I don't know if I've done the right thing.' And he said, 'Absolutely. I tell all my patients to take a notebook with them and write down everything that they suddenly think of that's bad. Face up to it.' He also told me he'd never heard of anyone with so many different personalities."

Anne did what Jane suggested. "And now I don't need a magnifying glass to read her writing," Jane said. "She no longer rings me all the time. She's living with her mother and works at a school for young children from disadvantaged families. They love her. And she gets great comfort from her two cats. She got her sister's case reopened and even braced herself to make public appearances on behalf of those who know the pain of a missing loved one."

I felt moved and inspired by how this young woman had been healing the trauma of her past and how Jane had found time to help her even as she was traveling nonstop around the world.

"It wasn't only wanting to help her," Jane said, as though to dispel any impression that she was a Mother Teresa. "It was also being utterly fascinated by her story. I've always been captivated by the mind and its problems."

"Sounds like it was the naturalist in you," I said. "What did you learn from working with her?"

"Well, she was a wonderful example of how our indomitable spirit can fight the worst abuse and pain and create a whole person again."

Jane had said that hope was a survival trait, and now I was start-ing to understand why. Somehow Jane was able to give Anne hope, and that set her on the path to recovery. When we face adversity, it is hope that gives us the confidence to rally our indomitable spirit to overcome it.

It seemed like we were returning to our earliest conversations about hope—how resilience is linked to the belief that we can make a difference in our lives and the lives of others, how hope really gives us the will to not only heal ourselves but to make the world a better place.

"You know," Jane said suddenly, after a companionable silence, "I think one of the most important things in all of this is having a sup-port network—which can, by the way, include animals. Remember Anne's cats."

We Need Each Other

"Yes, that's clearly true," I said. "My research on resilience has shown me the importance of social support in times of trouble—how im-portant it is in helping people overcome depression and despair and finding hope again."

"Oh yes, and you've made me think of a wonderful example," Jane said with a smile. I sat back in my seat to enjoy more story time.

"It's something I heard about during one of my visits to China. It's about two extraordinary men—hang on, I have to look up their names."

Jane opened her laptop. "Here we are—Jia Haixia and Jia Wenqi." Jane spelled the names for me and then closed her laptop and began a story she obviously loved.

"They live in a small village in rural China and have been friends since they were boys. Haixia was blind in one eye at birth from a cataract and lost sight in the other eye in a factory accident, and when Wenqi was only three years old he lost both his arms when he touched a downed power line. When Haixia lost his sight completely he became really depressed and Wenqi realized that he must find something they could do that would give purpose to Haixia's life. At that point they were in their mid-thirties.

"I don't know how long it took Wenqi to think up his plan, but he suddenly got the answer. Both had often talked about how the land around their village had become increasingly degraded since they were young. Quarrying had polluted the rivers, killing fish and other aquatic life, and industrial emissions had polluted the air.

"I can just imagine Wenqi telling his friend that what they should do was plant trees. And I bet Haixia was incredulous at first—how could they do that? They didn't have any money, and he was blind and Wenqi had no arms. Wenqi had the answer—he would be Haixia's eyes and Haixia would be his arms.

"They couldn't afford to buy seeds or saplings, so they decided to clone from branches cut from the trees. Haixia did the cutting while Wenqui directed him to the right place. And they walk from place to place with Haixia holding on to one of Wenqui's empty sleeves. At first it all went wrong. They were excited they managed to plant about eight hundred cuttings in the first year, but imagine how they felt when spring came and only two of them were alive. The land was simply too dry. At that point Haixia wanted to give up, but Wenqi told him that was not an option—they would just have to find a way to get water to the trees.

"I don't know how they did that—but, anyway, they did. They planted more cuttings, and this time most of them survived."

Jane said together they have now planted over ten thousand trees. At first the other villagers were skeptical, she told me, but now they help to take care of those very special trees.

"A documentary was made about them," Jane said, "and in it I remember Wenqi saying that if they worked together physically, and united spiritually, they could achieve anything. And he said—wait a moment." Jane opened her laptop again. "Yes, here it is. 'Though we are limited physically, our spirit is limitless. So let the generation after us, and everyone else, see what two handicapped individuals have accomplished. Even after we're gone, they will see that a blind man and an armless man have left them a forest.'

"And that," said Jane, "is a wonderful example of how friendship

A story from rural China: together, Jia Haixia and Jia Wenqi have planted more than ten thousand trees to help heal the degraded and polluted land surrounding their village—a blind man and an armless man. Talk about indomitable spirits. (XINHUA NEWS AGENCY REPORTER, CHINA GLOBAL PHOTO COLLECTION)

can give hope to the hopeless. And a marvelous illustration of what can be accomplished by the indomitable human spirit."

"So what you're saying is that the one person who is determined and sees where to go inspires others so that people then work on a problem together?"

"Yes," Jane replied. "And the other thing that is really important is to help people realize that they, as individuals, matter. That they each have a role to play. That they were born for a reason."

"And a sense of meaning is so important for hope and happiness, isn't it?" I asked.

"It is," Jane replied. "Without meaning, life is empty and day will follow day, month will follow month, and year will follow year in mindless succession."

"Those," I reflected, "are the people who have lost hope."

"And sometimes it is possible to rouse them from a seemingly meaningless life with a really good story," Jane said, "one that will reach their hearts and wake them up."

"Can you give me an example?

"One of my very favorites is fictitious but seems so appropriate now. It is *Lord of the Rings*."

"What makes it such an appropriate story for the hopeless?" I asked.

"Because the might the heroes were up against seemed utterly invincible—the might of Mordor, the orcs, and the Black Riders on horses and then on those huge flying beasts. And Samwise and Frodo, two little hobbits, traveling into the heart of danger on their own."

"Is that an example of the indomitable *hobbit* spirit?"

Jane laughed. "I think it provides us with a blueprint of how we survive and turn around climate change and loss of biodiversity, poverty, racism, discrimination, greed, and corruption. The Dark Lord of Mordor and the Black Riders symbolize all the wickedness we have to fight. The fellowship of the Ring includes all those who are fighting the good fight—we have to work so hard to grow the fellowship around the world."

Jane pointed out that the land of Middle-earth was polluted by the destructive industry of that world in the same way that our environment is devastated today. And she reminded me that Lady Galadriel had given Sam a little box of earth from her orchard.

"Do you remember how he used that gift when he surveyed the devastated landscape after the Dark Lord was finally defeated? He started sprinkling little pinches of the earth all around the country—and everywhere nature sprang back to life. Well, that earth represents all the projects people are doing to restore habitats on planet Earth."

I found Jane's metaphor both soothing and inspiring as I let myself imagine all the small and often humble ways people everywhere were doing their part to repair the harm we've caused. The fire had burned low, but the room and Jane's face were still illuminated by the now setting sun. It seemed a fitting image to close our conversation on—at least for this visit.

We still had one last conversation about hope I wanted to explore with Jane—one that had long interested me. I wanted to know about Jane's journey to becoming a global icon. How did she transform into a global messenger of hope?

But this last conversation about Jane's personal journey would have to wait until our next visit. We made a plan to meet again in a few months' time when I could talk with Jane in her childhood home in Bournemouth—which seemed ideal since I wanted to learn about her early formative years. When we hugged goodbye and I left the cabin at sundown, it was December of 2019. Little did we know when we parted ways in the Netherlands how interrupted our conversation about hope would become. And how even more urgently a conversation about hope would be needed.

People have always commented on my eyes, and said I looked as though I had ancient wisdom; an "old soul" was how one woman put it. (JANE GOODALL INSTITUTE/COURTESY OF MY UNCLE, ERIC JOSEPH)

III

Becoming a Messenger of Hope

A LIFELONG JOURNEY

Like so many other meetings, celebrations, and reunions around the globe, our plan for me to visit Jane in her childhood home in Bournemouth had to be canceled because of the pandemic. It wasn't until the fall of 2020 when Jane and I were able to resume our conversation. We spoke on Zoom—Jane was indeed in Bournemouth, but I was sitting on the other side of the world in my home in California.

By now the virus had caused an enormous amount of economic and emotional hardship and left death and devastation in its wake. Just days earlier, I'd attended the funeral of my college roommate. At the beginning of the pandemic he'd lost his job and become depressed. Another college friend and I were trying to support him through his disorientation and loss, when we finally discovered how despondent he'd become. He'd seemed to be doing better and told us he didn't need us to come to him or send him help. But two days after our last conversation, he shot himself.

My grief for my dear friend was part of a rising global trend, deaths of despair escalating in terrifying ways as people struggled with the dislocation and isolation that the pandemic had caused. Within months, another person close to me, a young family friend, would die of a drug overdose. A mental health pandemic was spreading as rapidly as the virus. So many people felt like they were being buffeted daily by a new crisis and waves of heartbreak and grief.

Seeing Jane's face, albeit on a screen, was a warm ray of hope in the midst of my grief. Her gray hair was pulled back into her typical ponytail, and she was dressed in the same green safari shirt that she had worn in Tanzania. She looked like a wilderness guide; and, indeed, during our work on the book, she had taken me to many of the most beautiful aspirations and darkest fears of our world and of our own human nature, as we were tracking hope and confronting despair.

"It's so wonderful to see your face after attending this brutal funeral," was the first thing I said when we reconnected.

"I'm so sorry, Doug. Losing someone we love is always hard. But suicide is an especially painful loss."

Jane was seated at a makeshift "desk"—a small box on a slightly bigger box on a tiny table. The shelves behind her were a collage of family photographs, mementos from her travels, and many of the books she'd read as a child, including the Doctor Dolittle and Tarzan stories, *The Jungle Book* about Mowgli being raised by the wild animals of India. And her collection of philosophers and poets, reminders of her adolescent curiosity.

"And I'm sorry you couldn't visit," Jane said, "but let me show you round my little hideaway up in the attic."

She walked the laptop around the room to introduce me to some of the people and keepsakes that mattered most to her.

"This is Mum," Jane said, as she picked up a framed photo of her mother with dark hair, wearing a brown shirt. "And this one is Grub," she said, pointing out a photo of her son, "he's about eighteen years

Mum. (MICHAEL NEUGEBAUER/WWW.MINEPHOTO.COM)

old here." Grub's hair was short, and he seemed to look forward through rimless glasses into his future.

"And here is Uncle Eric." He had dark hair and a serious, penetrating gaze. I was now seeing the likenesses of all the relatives I had met already during our conversations. "This is my grandmother Danny," she said, pointing out a large black-and-white photo of an elderly woman with a gentle face that was both determined and wise. And there was another of Danny with Grub when he was a three-year-old child. Next there was her aunt, known to everyone as Olly, a shortening of her Welsh name, Olwen. And a framed portrait of the grandfather Jane had never known, who had died before she was born, a serious yet warm face rising from his clerical dog collar. Finally, there were photos of both her husbands, Hugo and Derek, and a large framed one of Louis Leakey.

Rusty—my teacher. (JANE GOODALL INSTI-
TUTE/COURTESY OF THE GOODALL FAMILY)

Jane's photo collection was filled with animals as well as people.
"This," she said, with a new softness in her voice, "is Rusty." She
pointed to a photo showing adolescent Jane in riding clothes with
a black dog with a white patch on his chest sitting close beside her.
"But let me show you his portrait." She brought the photo close to
the laptop screen and I could see his clear eyes, full of intelligence.

"He was so special," Jane said. "More intelligent than any other
dog I've known. He's the one that taught me that animals have minds
capable of solving problems, as well as emotions and very definite
personalities, which of course helped me enormously when I began
studying the chimpanzees.

"And here is David Greybeard." I could see the distinguished

One of the most amazing experiences: Wounda embraced me for so long and I only met her that day. (JANE GOODALL INSTITUTE/ FERNANDO TURMO)

white chin hair of the first chimpanzee to lose his fear of her. The one who demonstrated to Jane that it wasn't only humans who used and made tools.

"And Wounda," Jane added.

I recognized the image from the video of this tender interspecies embrace that had gone viral. Wounda had been stolen from her home by poachers for bushmeat, and when she was rescued by one of the JGI chimpanzee rehabilitation centers, she was clinging to life. After the first chimp-to-chimp blood transfusion in Africa, she was nursed back to health and was taken to a protected forested island in the Republic of Congo. After she emerged from the traveling cage,

Wounda before and after. (JANE GOODALL INSTITUTE/ FERNANDO TURMO)

she turned and gave Jane a long embrace. Jane said it was one of the most amazing experiences she's ever had. Wounda has since become the alpha female and given birth to an infant named Hope.

"And up there," Jane said, as she angled her laptop, "are some special stuffed animals. I get given stuffed animals everywhere I go— mostly chimpanzees, of course!" She took down one of a black robin,

Some of the stuffed animals I've been given as I travel around the world.
(JANE GOODALL INSTITUTE/JANE GOODALL)

the species miraculously saved from extinction, which she had told me about in one of our interviews. She pointed out a few other toys representing endangered species that people are working to save.

Then, from a chair beside her desk, she picked up a strange-looking monkey holding a banana. I recognized him at once: the famous Mr. H.

"He was given to me by Gary Haun twenty-five years ago," Jane said. "Gary was blinded during an accident when he was twenty-one years old in the Marines. For some reason he decided he wanted to become a magician. 'You can't be a magician if you're blind!' people told him. But in fact, he's so good that the children, for whom he does shows, don't realize he's blind. When his act is over, he explains that he's blind, and tells them that if things go wrong for them, they must never give up, that there's always a way forward. He does scuba diving and skydiving, and

Gary Haun, the blind magician who gave me
Mr. H. He calls himself The Amazing Haun-
dini! (ROGER KYLER)

he has actually taught himself to paint." Jane picked up a book, *Blind Art-*
ist, and opened it to a picture of Mr. H. I was amazed that it was painted
by a man who had never seen, but had only touched, the stuffed animal.

"Gary thought he was giving me a chimp," Jane added, "but I
made him hold the tail, and, of course, chimps don't have tails. 'Never
mind,' he said, 'take him where you go, and you'll know I'm with
you in spirit.' So Mr. H has now been with me to sixty-one countries,
and he's been touched by at least two million people because I tell
them the inspiration will rub off on them. And let me share a secret
that I tell the children: every night Mr. H eats a banana, but it's a
magic banana and is always there again in the morning." She flashed
that mischievous, knowing smile.

Mr. H is very famous. Everyone, especially children, wants to touch him.
(ROBERT RATZER)

She picked up four more toys. "Let me introduce you to Piglet, Cow, Ratty, and Octavia the Octopus. Along with Mr. H they're also my traveling companions."

I asked her why they were special. "They illustrate points in my lectures. I use Cow when I'm talking about factory farming. Especially when I talk to children and I want to explain how they produce methane, that virulent greenhouse gas." She laughed, and holding up Cow she demonstrated. "Food goes in here"—she pointed at the mouth—"and while it's being digested, it creates this methane." She lifted Cow's tail to show where the gas emerged. "And I tell them cows burp, too. And there's lots of giggling. Ratty I use when I talk about how intelligent rats are, and especially how the African giant forest rat, or pouched rat, has been trained to detect land mines that are still active, left behind after a civil war." I knew how many people had lost a foot or leg from stepping on one of these. Jane told me that Piglet and Octavia were also props she used when she talked about animal intelligence—especially in pigs and octopuses.

She also pointed out various mementos of her life "on the road" when, before the pandemic, she traveled around the world spreading awareness—and hope. "Precious gifts, each with a story," she said, maneuvering her laptop camera lens slowly across the packed shelves.

Last of all she came to a wooden box with two chimpanzees exquisitely engraved on the lid. "In here," she said, "I keep most of my symbols of hope. Sometimes I use them in my lectures."

She returned to her studio, put the laptop back on her precarious desk, and one by one picked up various small items to show me. First was a clumsily made bell that gave off a somewhat unmusical ringing when she shook it. "This was made from metal from one of the many land mines that had remained buried and unexploded after the civil war in Mozambique. Hundreds of women and children lost a foot after treading on one when working in the fields. What makes it even more special is that it was detected by a specially trained African giant pouched rat, like the one I just told you about. They are adorable little creatures—I have watched them being trained in Tanzania and they are still working in different parts of Africa."

Next came a piece of fabric. While supervising the clearing of land mines for a charity in Mozambique, Chris Moon was blown up. He lost his lower right leg and lower right arm. Not only did he learn to run with his specially designed lightweight prosthetic, but he completed the London Marathon less than a year after leaving the hospital and subsequently entered many other marathons. "And this is the foot of one of the socks that Chris used to pull over his stump to try to

One of my symbols of hope, a bell made from metal from a defused land mine. I always ring it at the UN on International Day of Peace. (MARK MAGLIO)

prevent chafing," Jane told me. "A very special one that he used when he was running in the toughest marathon in the world, Marathon des Sables. And he completed the whole of the one hundred thirty-seven miles—across the Sahara Desert."

Jane then held up a piece of concrete that had been broken off by a German friend—who only had his penknife—on the night the Berlin Wall came down. She also had a piece of limestone from the quarry Nelson Mandela had been forced to work for while he was in the prison on Robben Island.

"And this is really, really special," Jane said, picking up a small greeting card. She opened it for me to see—inside were two very tiny black primary feathers that had been sent to her by Don Merton. We have already told the story of how he had saved the Chatham Island black robin from extinction. "These," said Jane, pointing lovingly at

Wildlife
Conservation

The California condor almost became extinct. Thanks to dedicated biol-
ogists, their numbers have increased. I love slowly extracting one of their
long primary feathers from its tube during my lectures. It is one of my
symbols of hope. (RON HENGGELER)

the little feathers, "are from Baby Blue, daughter of Old Blue and
Yellow."

She told me she also had a primary feather from the wing of a
California condor, another bird saved from extinction, but that it was
in the JGI office in America. Jane told me it was twenty-six inches
long! "I pull it very, very slowly from its cardboard tube when I'm
giving a lecture in the United States. It never fails to elicit a gasp of
amazement—and I think a sense of reverence."

Carefully Jane put her treasures back in their box. We resumed
our interview, and once more I was looking into those probing eyes. I

made some remark about them, and she smiled as a memory jumped into her mind. "When I was a baby—perhaps about one year old, my nanny used to push me around a park in my pram. Apparently, many people would stop to greet us—everyone knew everyone back then. But there was one elderly woman and she refused to look at me. 'It's her eyes,' she told Nanny. 'She looks as though she can see into my mind. There is an old soul in that child, and I find it disturbing. I don't want to look anymore.'

"Oh, hang on a moment," Jane said, suddenly stepping away from the screen. "I forgot to plug my laptop in—I'm about to run out of juice." As she fetched her power cord many thoughts came to mind. There were so many reasons to worry that our best days as a species were behind us. Political turmoil and the rise of demagogues threatened democracy around the world. Inequality, injustice, and oppression still plagued us. Even our planetary home was in peril. But despite it all, Jane had shown me some profound reasons for hope. In our amazing intellect, in the resilience of nature, in the energy and commitment of today's youth. And, of course, there is the indomitable human spirit. What was it about Jane that had enabled her to experience and agonize over so much cruelty and suffering of people and animals around the world, and so much destruction of nature, and yet remain a beacon of hope? Had this capacity been in her from the very beginning?

As Jane sat back down, I told her how amazed I was by her ability not only to have hope for the future but to inspire hope in others. "How did the infant in the pram with an old soul looking out of her eyes get to be this messenger of hope?" I asked her.

"Well, I think some of the answers to that question did begin to take shape when I was just a child," she said. "I've already talked about the self-confidence that I was given by my supportive mother. And growing up with such a wonderful family around me. Danny had to cope with her family when my grandfather died of cancer and left her almost penniless. Pity there's no time to share her story now. Then Olly and Uncle Eric were also wonderful role models. Olly was a physiotherapist who worked with many child victims of polio, club-foot, rickets, and so on; and my first job when I got home after my secretarial course in London was to take down notes from the ortho-pedic surgeon who came to examine the children once a week. And there I learned how cruel life can be, inflicting such grievous afflictions on innocent children and their families. And, too, I was again and again impressed by their courage, their stoicism. Hardly a day goes by when I do not give thanks for the gift of good health. I do not take it for granted."

She told me that Uncle Eric had stories of courage to share when he came to Bournemouth for weekends after operating on victims from the Blitz. "And as I've said," Jane said, "growing up in World War Two taught me so much—I learned the value of food and clothes because everything was rationed. And I learned about death and the harsh realities of human nature—love, compassion, courage on the one hand; brutality and unbelievable cruelty on the other. This dark side was strikingly revealed to me at a young age, as we discussed earlier, when the first reports and photos of the skeletal survivors of the Holocaust were published.

"And the defeat of Nazi Germany—well, there could be no better example of how victory can be won, even when defeat seems inev-itable, if the enemy is confronted boldly and with great courage."

I was beginning to understand the important role Jane's family

With my father, Mum, and Judy the day I was honored with a CBE (Commander of the British Empire). (JANE GOODALL INSTITUTE/MARY LEWIS)

and circumstances had played in creating the Jane of today. But I realized she hadn't mentioned her father.

"No, my father doesn't figure largely in my childhood memories as he joined the army in the Royal Engineers at the very start of the war and he and Mum divorced at the end of the war. But I surely inherited my very tough constitution from him."

"Yes, you've told me that you recovered from some really bad bouts of malaria and that the various bruises and cuts you received when climbing around in the forest always healed quickly. How did you become so strong, because you said you weren't that way in early childhood?"

Jane laughed. "I absolutely wasn't! I missed a lot of school. As I think I mentioned to you, I used to get really, really bad migraines—they tended to come on just as I started an end-of-term exam—which really upset me, as I always studied hard and actually looked forward to answering whatever questions were thrown at me. And I frequently had painful bouts of tonsillitis, which were several times accompanied by quinsies."

"What is a quinsy?" I asked.

"It's an abscess around the base of a tonsil—really, really painful until it bursts. And I got all the childhood diseases except mumps—measles, German measles, chicken pox—and Judy and I both nearly died of scarlet fever.

"And I told you about the time, when I was about fifteen, I became convinced that when I shook my head, I could hear my brain shifting inside my skull. I was really scared. In the end Uncle Eric had me examined. Of course, there was nothing wrong with my brain at all, but I still didn't dare shake my head because I still heard, or thought I heard, my brain moving around inside. In fact, as I told you during one of our talks, I was so often sick that Uncle Eric used to call me Weary Willy. But then, one day I overheard him talking with Mum and questioning whether I had the physical stamina to follow my dream to go to Africa. And that, of course, was like a challenge—if I wanted to fulfill my dream of studying animals in Africa, I had to prove him wrong!"

"And you certainly did. But how did you do that?"

"Thinking back, I realize that I was never sick during the holidays. I did really well at school, but I wanted to be out in nature. Getting sick must have been some kind of psychological—and totally unconscious—way of getting out of school! Because over the holidays I was a real tomboy, climbing the tallest trees, swimming when there was snow on the ground, allowed to ride the most spirited horse who loved to buck and try to run away at the riding school."

I laughed. "Perhaps all of that trained you for everything you had to contend with in Africa?"

"There were some frightening moments," Jane said. She told me of her close encounters with buffaloes and leopards—she came upon them suddenly, but they never harmed her. Once, a deadly Storm's water cobra was washed onto her foot as she was walking along the beach and stared up at her "from black expressionless eyes." She laughed. "I was a bit scared then, I must admit. There was no anti-venom, and so many fishermen had died from cobra bites when they accidentally caught one of those snakes in their nets. I just stood motionless—and was very relieved when another wave washed it away!

"But all of that was exciting, Doug," Jane said. "The worst part was when the chimps were running away from me and I was not sure I had time to win their trust before my funding ran out. People have asked me if I ever felt like giving up in the beginning. Well, you know me too well by now—I'm obstinate and I never even thought of quitting."

"What about when your methods were criticized when you got to Cambridge University?" I asked. "After all, you had not been to college. You'd had no scientific training. Weren't you intimidated then?"

"I was intimidated at the thought of going to that prestigious university and being among students who had had to work so hard for their undergraduate degrees. But when I was told I could not talk about chimpanzees having personalities, minds, and emotions—well, I was just shocked. It was lucky I had learned from Rusty, and the various pets I had as a child even before I got to know the chimpanzees, that in this respect the professors were absolutely wrong.

I knew very well that we are not the only beings on the planet with personalities, minds, and feelings, that we are part of, and not separate from, the amazing animal kingdom."

"So how did you tackle those professors?"

"Well, I didn't argue—I just sort of quietly went on writing about the chimps as they are, showing the film that Hugo had taken at Gombe, inviting my supervisor to Gombe. What with all my firsthand observations along with Hugo's fantastic film, and the facts about their biological similarity to us that had emerged—well, most scientists gradually stopped criticizing my unorthodox attitude. Again, I'm pretty obstinate and I don't give up easily!"

I thought about that victory—which is now considered to have played a key role in changing our relationship with animals.

"Anyway"—Jane interrupted my thoughts—"as you know, I got my Ph.D. and went back to Gombe and would happily have stayed there forever, but of course that all changed when I attended that 1986 conference and had my Damascus moment."

"What happened after that?" I asked.

"Well, the first thing I decided to tackle was the nightmare of chimpanzees in medical research."

"Jane," I said, "did you really think you could do anything to help the chimps in those labs? Did you really think you could stand up to the medical research establishment?"

Jane laughed. "Probably if I'd really thought it through I would never have tried. But having seen those videos of the chimps in the labs—well, I was so upset and angry that I just knew I had to try. For the sake of the chimps.

"The worst part was forcing myself to actually go into a lab to see what it was like with my own eyes. I don't think you can tackle

(Top) A juvenile in a research lab, far gone in depression. Note the size of the cage. (LINDA KOEBNER) (Bottom) Me visiting a chimp in one of the lab prisons. (SUSAN FARLEY)

any problem without some firsthand knowledge. Goodness, how I dreaded being in the presence of intelligent social beings who were confined, alone, in five-foot-by-five-foot cages. In the end I went to several labs, but my first visit was the hardest. Mum knew how I felt, and she sent a letter enclosing a card on which she had written a couple of Churchill's quotes. And, amazingly, driving to the lab we passed the British Embassy with the statue of Churchill making his famous V for Victory sign. It was like a message from the past. Once again that inspiring wartime leader was there to give courage when I so desperately needed it."

"How did it go once you got there?"

"The visit was even more heartbreaking than I expected—and it made me even more determined to do all I could to help those poor prisoners," Jane said. "I decided to use similar tactics to those I used with the Cambridge scientists—I talked about the behavior of the Gombe chimps and showed them films. I truly believe that a lot of what I perceive as deliberate cruelty is based on ignorance. I wanted to touch their hearts—and for some of them, anyway, it worked. We had meetings; they invited me to give talks to their staff; and they agreed to at least allow me to send in a student to introduce 'enrich-ment' into some of the labs. Something to alleviate the desperate boredom of an intelligent being in a bare cell, alone and with nothing to help pass the monotonous days—except times of fear and pain resulting from invasive protocols.

"It's been a long, hard fight with many individuals and groups helping, but finally chimpanzee medical research, so far as I know, has ended. And though my fight was on ethical consideration, the final decision that affected the approximately four hundred chimpanzees owned by NIH in America was made when a team of scientists found that none of the work being done was truly beneficial to human health."

I was aware that this was the first of many battles Jane threw herself into over the years, but I asked her how she went on to attack the enormous challenges that faced her beloved chimpanzees in Africa.

Challenges in Africa

"So after many years you and others who joined the fight won that battle. But at the same time you were trying to do something about the situation in Africa, right? Wasn't that even more difficult? Did you really think you could make a difference?"

"Oh, Doug! I really didn't know if I could! It was after that conference in 1986, the one where I saw the secretly filmed footage of chimps in labs. I didn't see how I could help them, but, as I told you, I knew I had to try. And at the same conference we had a session on conservation—and it was shocking. Images from across Africa of forests destroyed, horror stories of chimps shot for bushmeat and infants snatched from their dead mothers to sell, and evidence of a drastic decline in numbers of chimpanzees wherever they were being studied. Again, I just knew I had to do something. I didn't know what or how—only that just doing nothing was not an option.

"And again I felt I had to go and see some of what was going on in Africa for myself. So I got enough money to visit six of the countries where wild chimpanzees were being studied. And one of the first challenges was the number of orphan infants whose mothers had been shot for bushmeat. Often the infants were sold in local markets as pets. It was illegal, but people had other problems to worry about—and corruption was rife.

"I shall never forget the first of these orphans that I saw. He was about a year and a half old, tied up on top of a tiny wire cage with a piece of rope. Surrounded by tall laughing Congolese. Curled up on his side, eyes blank, staring into nothing. But when I went close and gave the soft chimpanzee greeting sound, he sat up and reached a hand toward me, looking into my eyes.

"Again I knew I had to do *something*. And I'd had a lucky break. Just before I started my trip to Africa, I was invited to a private lunch by James Baker, when he was secretary of state under George Bush Sr. And he offered to help. He sent telex messages to all the American ambassadors of the countries I planned to visit, asking them to help me. So I was able to appeal to the ambassador in Kinshasa, and he spoke to the minister of the environment, who sent a policeman with us when we returned to the market that evening. It was deserted except for that one little chimp—I think word had gone around about the police! We cut the rope, and Little Jim, as we called him, in honor of the secretary of state, clung to me with his arms round my neck. Of course, I couldn't care for him, but he was transferred to the loving care of Graziella Cotman, the woman who had begged me to go to Kinshasa and see if I could help. That was the start of our sanctuary programs for orphan chimps.

"And we've already talked about how I realized that to help the wild chimpanzee situation it would be necessary to improve the lives of local communities, many of which were suffering the effects of extreme poverty—and how that led to Tacare."

From Shy Young Woman to Global Public Speaker

By this time I was beginning to understand how Jane had managed to tackle what many believed were insoluble problems—through determination and having the ability to inspire and get help from those in the best position to bring about change. But what about her transformation from a field researcher who spent hours alone in the forest to someone traveling and giving talks three hundred days a year, always surrounded by people?

"What enabled you to make that transition?" I asked her. "You've told me you were a shy child—how would you have felt if someone had prophesied to your twenty-six-year-old self what your future would be?"

"If someone had told me when I first went to Africa that at some point I would have to give lectures to large auditoriums full of people   well, I would have said that that would be impossible. I had never spoken in public. And when I was told that I had to give a talk   well, I was terrified.

"And for the first five minutes or so of my first lecture, I felt I couldn't breathe. But then I found it was okay. I could breathe again. And that is when I first realized that I had this gift. A gift of being able to communicate with people. To reach their hearts—with spoken as well as written words. Of course, I've worked hard to get better. When I was practicing with my poor family for that first talk, I made a vow: I would never read a speech. And I would never say 'um' or 'er.'"

"Why did you make that vow?"

"Because I thought people who read speeches were boring. And lots of *ums* and *ers* are irritating."

I loved hearing this legendary speaker describe the first speech she gave and how she had resolved to practice—to give it her all.

"Anyway, the gift was there waiting to be used, right from the start. I remember the third talk I ever had to do was at the Royal Institution in London, where lots of famous British scientists have spoken. The tradition was that no one introduces you—you simply walk to the podium as the clock is chiming eight, and as the last chime ends you start your lecture. And on the first stroke of nine, precisely, you must stop. So I'm terrified, completely terrified. I had to attend a small formal dinner beforehand and then they put me alone in a room for an hour."

"But isn't that what you always want?" I asked. "Time to be by yourself, to focus?"

"That's what I want now, but back then—well, it was just an hour to get more and more nervous! And as they led me to that room, I panicked as I realized I'd left my notes behind!

"Frantically I asked somebody to make a call to Mum, and she was able to come early and bring my notes to me. That calmed me down a bit. But I remember pacing round and round that little room."

I asked her how it went.

"Well—I was led out there, like a lamb to the slaughter. I walked out onto the platform. I remember the old clock made a whirring sound as it got ready to chime the hour. And—well, I began to talk at the last stroke of eight and I ended, *exactly* where I meant to end, on the first stroke of nine.

"Afterward, one of the staff asked me for the transcript of my talk, and I said, 'What do you mean?' And he said, looking surprised, 'Well, you know, what you read from.'

"He looked amazed and slightly bewildered when I handed him a single piece of paper on which there were about six or seven scribbled lines in red ink!"

"You have been giving talks to huge audiences for decades now," I said. "Did you have any sense at the time that this first public speech might lead to so many more?"

"Well, I always knew I had a gift for writing," Jane added. "From an early age I was writing—stories, essays, poems. But I never thought I had a gift for speaking. It wasn't until I was forced to make that first speech, and found that people were listening, and heard their applause at the end, that I realized I must have done okay. I think many people have gifts that they don't know about because nothing forces them to use them."

I thought about this for a moment, then asked Jane if she believed she had been given that gift for a reason.

"I sort of have to believe that," she said. "I know I have been given certain gifts, and it does seem there is a reason for that. In any case, whether or not there is a reason, I feel I simply must use them to do my bit to make the world a better place and for the good of future generations. And although it feels strange to admit it, even to myself, I do somehow believe I was put here for a reason. I mean, when I look back over my life I can't help thinking that there was some kind of path mapped out for me—I was given opportunities and I just had to make the right choices."

"Let's Just Say It Was a Mission"

"So you're this shy person and yet you signed up for a life of speaking—"

"I didn't sign up for it," Jane interrupted. "It overtook me. It swept me up in its path."

"Okay, it swept you up—but you agreed. You went along with it."

"I didn't have an option."

"You felt called?"

"I wouldn't put it quite like that—it was just—well, 'The chimps have given me so much, now it's my turn to try to do something for them.' People say it must've been a hard decision to make, leaving Gombe. But it wasn't. I've told you that. It was like Saint Paul on the road to Damascus. He didn't ask for that to happen. He didn't decide, according to the story, anyway. It just happened—he changed from persecuting the early Christians to trying to convert people to Christianity. It was a huge change—that's why it's the best example I can think of."

"So did that sense of being called—"

"No, let's just say it was a mission," Jane interrupted.

"Okay—did that sense of mission stop you from having self-doubt, or were there times when you said 'I don't know if I can do this speech or talk to this prime minister or CEO'?"

"Of course, there were. There still are. I remember when I was first asked to attend one of the big UN climate change conferences. It would take me right out of my comfort zone—which is speaking with a crowd of students or to an auditorium full of the general public. My friend Jeff Horowitz, who has worked tirelessly to protect forests to mitigate global warming, asked me to sit on a panel with climate experts and CEOs of huge corporations and government representatives. I just said, 'I can't do that, Jeff. Honestly, I can't.'"

"What made you feel you couldn't do it?" I asked.

"Because I'm not a climate scientist. But Jeff did not accept my refusal, and in the end, I thought, 'Well, if Jeff believes in me, and feels it will help, I will just have to do my best.' Of course, I know now that people want to hear someone speaking truthfully about what we are doing wrong, especially when they can be reassured that there is a way out of the mess we have made. They want to hear people who speak from the heart. They want to be given reasons for hope. But even knowing that, I still get nervous."

It was amazing—and maybe encouraging—to hear the self-doubts of one of the world's most famous conservationists. I also thought that if it was true Jane was born for a reason, she was given a very difficult path to follow, with many problems to tackle. But I was beginning to see that once Jane decides on a course of action, nothing will stop her. Indeed she, too, had an indomitable spirit.

"You've had so many challenges and overcome them," I said to Jane. "You've said you're obstinate, and that you won't give up. And it's clear you have certain gifts that help, especially being able to reach people's hearts. Is there anything else that's helped you become a messenger of hope?"

"Yes, I've been so lucky—I've always had amazing people supporting me. I could never have achieved the things I've achieved on my own. It started with my family, of course, and then I've somehow been able to persuade so many people to help. One person has been there for me, someone with whom I can share feelings of sadness and anger: Mary Lewis. She's been working with me for thirty years. And Anthony Collins has helped in Africa, as a wise friend and counselor for just as long. But wherever I go, there's always someone to nurture me, reach out a helping hand, share a meal and a laugh. Oh—and a whisky, of course! I could not have done what I've done without them all. We've succeeded together."

I thought of our last conversation—about the importance of social support during difficult times.

"Another thing that has helped me to face so many daunting challenges is my grandmother's favorite text from the Bible: 'As your days, so shall your strength be.' When I'm lying awake the night before having to make one of those speeches, I say that to myself. It reassures me."

"What does that mean to you, that text?"

"That when the trials of life come, you'll be given the strength to cope with them, day by day. So often I've thought at the start of a dreaded day—having to defend my Ph.D. thesis, giving a talk to an intimidating audience, or even just going to the dentist!—'Well, of course, I shall get through this because I have to. I will find the strength. And, anyway, by this time tomorrow it will be over.'

"And there's something else, too. When I'm feeling the most desperate—when I'm so tired, so completely exhausted, that I feel I absolutely cannot give a lecture, I've somehow found a hidden strength that lets me cope with what's demanded of me."

I asked her where that hidden strength came from and how she found it.

"I sort of open up my mind to some kind of outside force," Jane said. "I just relax and decide to appeal to the source of the hidden strength, to that spiritual power that seems to have sent me on this mission; and in my mind I say, 'Well, you got me into this horrible situation, so I'm counting on you to get me through.' And it seems that on those occasions I give some of my best lectures! It's strange—once or twice it's been as though I've actually been able to see myself from outside giving those talks."

"You looked up," I said, "when you talked of an outside force."

"Well, it's not down there," Jane said, pointing to the ground and grinning.

"So you just empty your mind and somehow trust that whatever that spiritual power is, it will somehow get you through the talk?" I asked. "And then in a sense you become a channel—you open yourself up to a wisdom that's greater than your own?"

"Well, yes, for sure. There's a wisdom that's far, far, far greater than my own. I was so thrilled when I found that the great scientist Albert Einstein, one of the most brilliant minds of the twentieth century, came to the same conclusion based on pure science. He said it was the harmony of natural law—it's a terrific quote, actually."

I started to respond but noticed Jane had suddenly looked away with a worried expression on her face. "Doug, I am sorry to interrupt here, but I see the robin on my bird table looking in through the window. He'll be angry if I don't feed him!"

"A bird table?" I asked.

"It's a little platform attached to the sill of my attic bedroom window," Jane said, still looking away to her left.

"Why don't you see if you can google that Einstein quote while I feed him. It's in his book *The World As I See It*."

While Jane was gone, I looked it up. And there it was, in the book Jane suggested: "The harmony of natural law . . . reveals an intelligence of such superiority that, compared with it, all the systematic thinking and acting of human beings is an utterly insignificant reflection."

I thought about this quote in the context of all we had talked about that day, and it occurred to me that as Jane followed her extraordinary path through life there must have been either some lucky coincidences, or she could have been guided by this superior intelligence that Einstein believed in. When Jane returned, I read out the full quote and then asked her, "So do you think you're being guided by this superior intelligence, or do you think good old coincidence plays a role in directing your journey—all of our journeys?"

Was It Coincidence?

"I simply cannot believe in coincidences, not anymore," Jane said without hesitation.

"Why not?"

"Well, a coincidence implies a random event occurring in juxta-position with something happening in your life—and I can't believe that all of the seeming coincidences in our lives are random. It's more as if they are offering us opportunities. I've had so many strange experiences."

"Like what?"

"One of them saved my life. It was during the war and Mum had taken Judy and me for a holiday—very close to home but where there was a stretch of beach where we could actually paddle as there was a small gap in the barbed wire defenses. We were staying in a little guest house—lunch was at twelve o'clock and if you were late, too bad. No lunch. On this one day Mum insisted on walking back a very long way round, involving crossing some sand dunes and going through a little wood. Which meant, we complained, that we would miss lunch. But she was adamant, so reluctantly we had to agree.

"When we were halfway back I distinctly remember looking up into the very blue sky and seeing a plane high, high up. And as I watched, two black cigar-shaped objects fell, one from each side of the plane. Mum urgently told us to lie flat in the sand and she sort of lay on top of us. Next thing there were two terrific explosions. It was very frightening. And afterward we saw that one of the bombs had made a great crater right in the middle of the lane—just where we would have been had we taken the route we had taken every other day.

"So—was it 'coincidence' that Mum decided to go that way? She had a murmuring valve in her heart and always avoided long walks."

"Did she tell you what prompted her to take that route?" I asked.

"No, she never liked to talk about it. But it was as though she had a sixth sense. There was another time when she crossed London during the Blitz to get her sister Olly out of hospital—both her legs were in a plaster cast after some operation. Mum had a terrible job to get her back to Bournemouth through war-torn Britain. Everyone thought she was crazy. The next day a bomb fell on the hospital—or maybe it was a nursing home. There is no one I can ask now."

"Can you explain this sixth sense?"

"Not really—it seems like mind reaching out to mind. Maybe Mum sensed the presence of the German pilots in that bomber, or had a premonition of the bomb that could have killed Olly. And there was another thing. She was very fond of my father's brother, and one evening, here in Bournemouth when she was having a bath, she suddenly screamed out his name and then started crying. Later she found it was the exact time when his plane was shot down and he was killed."

I wondered why Jane's mother did not like to talk about her sixth sense, and Jane said she found it spooky.

"I've got another story about this kind of coincidence. It was when Grub was in boarding school in England, on the night my husband, Derek, died, far away in Tanzania.

"Grub had the same kind of premonition. It came to him in a

strange way. He woke suddenly from a dream in which Olly had arrived at the school and said, 'Grub, I have something very sad to tell you. Derek died last night.' He had the dream three times, and the third time he went to the school matron to tell her that he was having terrible nightmares. In the morning, Olly arrived at the school and took him into the garden. She said, 'Grub, I have something very sad to tell you.' Grub said, 'I know, Derek is dead.'"

As I thought about these stories I realized we had left the realm of science, but I was still intrigued.

"I want to tell you about another 'coincidence' that made a difference in my life. There was one empty seat on a Swissair flight from Zurich to London. I should have been on a later flight, but my plane from Tanzania had arrived early and I had changed to an earlier flight. The only empty seat on the whole plane was the one next to me. The man who took that seat arrived just before the doors closed. He told me he should have been on an earlier flight, but his connecting flight had arrived late. He seemed busy—I did not speak, apart from our polite greeting, until toward the end of dinner when I started a conversation. I was on my way—very inexperienced and scared, mind you—to do a TV interview with the head of a powerful pharmaceutical company, Immuno, that was using chimps for HIV research in their Austrian lab. They had filed seventy-one lawsuits against seventy-one people or groups who had challenged them over the conditions in their lab. This was 1987 and I was crazy or stupid enough to agree to this confrontation on TV. Well, it turned out my seat mate was Karsten Schmidt who was—I think—then the chairman of Baker and McKenzie. He told me not to worry—he would take up my case pro bono if they sued me! Subsequently, Karsten joined the board of JGI UK, drafted the statutes, and was our board chair for many years. Was it coincidence that put us next to each other on that plane that *neither of us* should have been on—and that we had the last two seats? If I had not initiated a conversation that opportunity would have been lost."

"Are you always on the lookout for opportunities?"

"Yes, even if I feel tired I always ask myself if perhaps there is a reason why I am sitting next to a particular person on a plane. Or at a conference. Anyway, it's worth a small effort just in case. And I've met some interesting people that way, some of whom have become friends and supporters."

"So you think you meet people for a reason?"

"Well, I don't really know. But I love to think about how things work out. Think of all the events and meetings that have led to the birth of every individual. Take Churchill. We start way back in the dim mists of the past when one individual man met one individual woman; they married; they had a daughter or a son and she or he met a man or a woman; and they had a child. And so it went on until all those meetings and couplings produced a Churchill."

"Or a Hitler," I said, a little skeptical about Jane's seeming belief in fate or destiny. I said as much to Jane.

"But I don't believe in fate or destiny. I believe in free choice," Jane countered. "Shakespeare put it so beautifully: 'The fault, dear Brutus, is not in our stars, but in ourselves, that we are underlings.' I believe that opportunities arise and you can seize them, reject them—or simply fail to notice them. If people had made different choices through the centuries, there would have been no Churchill and no Hitler."

"Nor a you or a me," I said.

I paused as I thought about it. It gave me a sense that I was part of a long lineage of love and heartbreak, longing and suffering that puts my own struggles into perspective. It helps me to feel that I am not alone and that I am not just living for myself. I am part of something greater than myself—but I don't know if it is all unfolding according to a plan.

"I am getting the feeling that your underlying belief in what you call 'a great spiritual power' is the source of a lot of your incredible energy and determination," I said. "How do you reconcile your spiritual orientation with your scientific mind?"

Spiritual Evolution

"When you talk about spirituality, many people are uneasy or absolutely put off. They think of a touchy-feely tree-hugging hippie sort of thing. Yet more and more people are now realizing that we have become increasingly materialistic and that we need to reconnect spiritually with the natural world. I agree—I think there is a yearning for something beyond thoughtless consumerism. In a way, our disconnect with nature is very dangerous. We feel we can control nature—we forget that, in the end, nature controls us."

Suddenly Jane said she'd just noticed it was 12:30 p.m.—the time when she took the old dog, a whippet called Bean, for his midday walk. "Of course, he has access to the garden," Jane said, "but he is a creature of habit. I won't be long. But I'll need to grab a biscuit and some coffee. Give me a thirty-minute break." And I was more than happy, as it gave me time to get some food, gather my thoughts and prepare the last questions.

Jane was true to her word and reappeared on her screen after exactly thirty minutes. I began the conversation, telling her that

I wanted to go back to the topic of our moral and spiritual development.

Jane immediately picked up the thread of our last exchange.

"Well, we as a species are on the road of moral evolution, discussing right and wrong, how we should behave as individuals toward each other and toward society, and our efforts to build democratic forms of government. And some people are also on the road to spiritual evolution."

"What is the difference between moral and spiritual evolution?" I asked.

"Moral evolution, I think, is understanding how we should behave, how we should treat others, understanding justice, understanding the need for a more equitable society. Spiritual evolution is more about meditating on the mystery of creation and the Creator, asking who we are and why we are here and understanding how we are part of the amazing natural world—again Shakespeare says it beautifully when he talks of seeing 'books in the running brooks, sermons in stones, and good in everything.' I get a sense of all of this when I stand transfixed, filled with wonder and awe at some glorious sunset, or the sun shining through the forest canopy while a bird sings, or when I lie on my back in some quiet place and look up and up and up into the heavens as the stars gradually emerge from the fading of day's light."

I sensed that Jane was lost in the beauty of the experiences she was describing. When she looked back at me again, I asked her whether she thought chimpanzees ever had similar feelings.

"When food is plentiful and the chimps have fed well and are content, they certainly have time for thinking. When I watch them gazing up through the canopy, or lying in a comfortable nest ready for the night, I'm always wondering what they are thinking, freed for a moment of planning where they will go for their next meal. And I do think it's possible that they have a similar sense of wonder, of awe. If so, it could be a very pure kind of spirituality—or at least a precursor of the kind of spirituality that we are talking about that is free from words.

"At Gombe there is a glorious waterfall, the Kakombe Falls, where the small stream plunges eighty-three feet, twenty-five meters, down a vertical groove worn by the falling water into the hard gray rocks of the cliff. There was the roaring sound of the falls on the rocky, gravely streambed, the constant breeze caused by the air displaced by the falling water. Sometimes as a group of chimpanzees approaches the falls, their hair bristles with excitement and they perform a wondrous display, standing upright and swaying from foot to foot, bending to pick up and throw rocks ahead of them into the stream, climbing the vines hanging down the rocks, and pushing out into the spray-laden breeze. After such a display, which may last at least ten minutes, they may sit gazing up at the water as it falls and watching it as it flows past them and away. Do they perhaps experience a similar emotion to the awe and wonder that I feel when I sit by that spectacular fall, listening to the thunder of the water as it crashes down onto the streambed?

"It always makes me realize the importance of our spoken language," Jane went on. "If the chimpanzees really do have this sense of awe, and if they could share this feeling with each other with words—do you see what a difference it could make? They might ask each other, 'What is this wonderful stuff that seems alive, that is always coming, always going, always here?' Don't you think that these questions could have led to the animistic religions, the worship of the waterfall, the rainbow, the moon, the stars?"

Kakombe Falls. (JANE GOODALL INSTITUTE/CHASE PICKERING)

"So you think that formal religions may have derived from those animistic religions," I said.

"I can't answer that, Doug. I'd need to be a scholar of religion, wouldn't I?"

"But you do believe in a spiritual power, a Creator—God—and that you were born into this world for a reason?"

"Well, it seems so. There are really only two ways in which to think about our existence on Earth. You either agree with Macbeth that life is nothing more than a 'tale told by an idiot full of sound and fury signifying nothing'—a sentiment echoed by some cynic who said that human existence is nothing more than an 'evolutionary goof.' Or you agree with Pierre Teilhard de Chardin when he said, 'We are spiritual beings having a human experience.'"

Although I often see myself as secular and not necessarily a believer of any particular religion, I was moved and inspired by what

Jane had said, and I was intrigued to explore the views of a scientist—watching my father die had raised questions that I wanted to try to answer. So I pressed Jane to tell me more about her beliefs.

"Well, I do not try to persuade anyone to believe, as I do, that there is Intelligence behind the creation of the universe, a spiritual force 'in which we live and move and have our being,' as the Bible puts it. I can't tell you why I believe this—I just do. And this is what truly gives me the courage to carry on. But there are many people leading ethical lives, working to help others, who are neither religious or spiritual. I'm just talking about my own beliefs."

Jane told me that many scientists, like Einstein, have come to the conclusion that there is "Intelligence" behind the universe. She said there have been more who proclaim themselves agnostic than atheist. Francis Collins, director of the National Institutes of Health, who led the team that worked on unraveling the human genome, began this work as an agnostic but was compelled to believe in God because of the awesome complexity of the information sent to every cell in the human embryo. Information that instructed it to develop into part of a brain, or foot, or kidney.

We discussed this for a while, and Jane confided that she truly welcomes this convergence of science and religion and spirituality.

"Because, you know, Doug, for some people I think that their religion is their only hope. Imagine if you have lost your whole family in war or some other disaster. That you are destitute. You arrive in a foreign country that agrees to take you in. You know no one. You cannot speak the language. What helps such people, I think, is if they have their faith. It is a firm belief in God—or Allah or whatever name they use—that gives them the strength to keep going. My wise mother told me that, as I was born into a Christian family, we talked about God, but that if we were a Muslim family, we'd worship Allah.

"She said there could only be one supreme being, the Creator 'maker of heaven and Earth' and it didn't really matter what name was used."

"So you think there is a heaven?"

Jane laughed. "Well, it depends on how we define heaven, I suppose. I don't believe in angels playing harps and all that sort of thing, but I'm sure there is something. Surely we shall see again those we have loved—definitely including animals! And also being able to understand the mysteries because we shall be *part of* them, part of the great pattern of things, but in an integrated way. I have experienced almost mystical moments of awareness when I have been alone in nature that might foreshadow the sort of heaven I like to imagine."

Little did I know this question about heaven would spark a final topic in our dialogue that was both profoundly mysterious and hopeful, especially as I was still grieving my father.

I had noticed that Jane sometimes had a slightly mischievous and knowing smile, like she had a secret. That was the smile I saw on her face now.

Jane's Next Great Adventure

"Last year, during question time at the end of one of my lectures, a woman asked me, 'What do you think your next great adventure will be?' I thought for a bit and then I suddenly realized what it might be: 'Dying,' I said.

"There was a *deathly* hush, a few nervous titters, and then I said, 'Well, when you die, there's either nothing, in which case, fine, or there's something. If there's something, which I believe, what greater adventure can there be than finding out what it is?'

"Afterward the woman came up to me and said, 'I never, ever wanted to think about dying, but thank you, because now I can think about it in a different way.' And since then I've mentioned this in several lectures and there is always a very positive reaction. And by the way, I always make it perfectly clear that it is just the way I think about death, and that I certainly don't expect everyone to feel the same way about it."

I thought back to my dad's illness and dying process, which was quite brutal as the cancer spread through his spine and brain.

"Do you think then that what people are scared of is illness—the process of dying—rather than actual death?" I asked.

"Oh yes," said Jane. "It is worrying about what will cause our death, what horrible illness or dementia or being bedridden and utterly dependent on others—those are what we all fear. But death itself is something else entirely. My grandmother Danny, at the age of ninety-seven, was more or less confined to her bed after living through bronchial pneumonia. One night Mum went up with a bedtime cup of tea and found her reading the letters from her late husband—she always called him Boxer—who had been gone for over fifty years. Danny smiled and said, 'I think you should write my obituary tonight, darling.' The next morning when Mum went in, Danny was lying peacefully on the bed. Dead. On her chest were all Boxer's letters, tied up in a red ribbon, with a note—'Please send these with me on my last journey.'"

We were silent for a moment or two and I could see Jane's eyes had teared up.

"Jane," I continued gently, still wanting to explore death and the adventure ahead, "does this mean you believe in reincarnation?"

"So many different religions do believe in it," Jane said thoughtfully. "The Buddhists believe that we can be reincarnated as animals—it depends on where we are on the path toward enlightenment. And, of course, both Hinduism and Buddhism believe in karma—if you suffer misfortune, you are paying for sins you did in a previous life.

"Honestly, I don't know—but I sort of feel that if there truly *is* a reason for our being here on this planet, then surely we wouldn't be given just one chance. When you think of eternity and our tiny little life span, it'd be awfully unfair! And you know," she said, grinning, "I sometimes think that what's happening in the world is just a test. Imagine Saint Peter at the gate to heaven getting out a computer printout of our time on Earth and checking to see if we used the gifts we were given at birth to try to do good!" Jane laughed.

I chuckled at Jane's image of Saint Peter as the examiner evaluating how we did on our Earth experiment. I thought of my dad's belief that life was a curriculum and also remembered the famous Jewish story about a Rabbi Zusha who was crying on his deathbed. When he was asked why he was crying, he said, "I know God is not going to ask me why I was not more like Moses or more like King David. He's going to ask me why I was not more like Zusha. Then what will I say?" I loved this story because it was a reminder that each of our curriculums is unique and each of us is meant to do our part in our own special way. It was clear that Jane had thought a lot about these things, and she clearly believed that death was not the end.

"You know, before my father died, he thanked me for companioning him on what he called 'his mighty journey to death,'" I said. "Like you, he definitely felt there was more to come." I told Jane how my son and I would FaceTime with my dad in his hospital bed when

I couldn't be with him. Jesse said he would miss speaking to my dad through FaceTime. My dad said not to worry, after he was gone we could speak through SpaceTime.

Jane laughed at his play on words. "There's that so important sense of humor in times of stress," she said.

"What do you say to people who think there is nothing more?" I asked.

"Well, first of all, as I told you, I never try to force my beliefs onto anyone else. But I tell them some of the amazing stories about near-death experiences. Elisabeth Kübler-Ross, who did so much research on this subject, writes about a woman who was pronounced brain dead on the operating table before being resuscitated. When she came to, she described the movements of people whom she could not possibly have seen from her position on the operating table. She described how she looked down as she sort of hovered over the room."

I told Jane about Bruce Greyson, who has been studying people who have had near-death experiences for forty years and has some pretty interesting stories of people who died and whose consciousness seemed to go on in some way, that consciousness itself seemed not to be limited to our brains.

"Once, when he was a young medical resident, he spilled spaghetti sauce on his tie in the cafeteria of the hospital," I said. "When he later went to attend to a young college student who had been brought to the hospital unconscious from a drug overdose, he didn't have time to change his tie, so he buttoned up his white lab coat to cover up the stain. Remarkably, when the patient regained consciousness, she told him she had seen him in the cafeteria and she described the stain on his tie. Yet the whole time that he was in the cafeteria, she had been unconscious in her bed, watched by a sitter.

"After that he studied many people who had seen and learned things during near-death experiences that should not have been possible, like meeting relatives that they did not know they had. He said that after their experience they almost universally believed that death was not something to fear and that life continued in some form beyond the grave. It also transformed how they lived their lives because they believed there is meaning and purpose in the universe."

I reminded Jane what she had just said jokingly, about life perhaps being a test, but that Greyson believed this could be true.

"He said that many of the people he'd talked to had experienced a sort of end-of-life review, where they literally saw their whole life flash before them," I said, "and that this had helped them understand why certain things had happened. Often they saw conflict from the other person's perspective or what caused people to act in the way they did. He talked about one truck driver, who had beaten up a drunken man who had cursed at him. When he had his near-death experience, he saw that the drunken man had recently lost his wife. Heartbroken, he had been driven to drink, which was why he had behaved in that abusive way."

"It's all absolutely fascinating, isn't it?" Jane said, her eyes alight with curiosity, a naturalist eager to explore virtually uncharted territory. "But unfortunately this adventure will have to wait till I'm dead.

"However, I do have some sort of proof," she added, "though it's not proof in the scientific sense—it is just an experience that proves it to me, and I don't care if anyone else believes it or not. It happened about three weeks after Derek died, when I was back in Gombe where Derek and Grub and I had had so much joy. I finally fell asleep listening to the waves and the crickets. Then I woke, or at least I thought I did, and I saw Derek standing there. He smiled and spoke to me for what seemed a long time. Then he disappeared and I felt I must quickly write down what he said, but even as I thought this, I felt a great roaring in my head as though I was fainting. I came out of this state, and once again I felt I must write down what I had learned, but once again the roaring, fainting feeling came over me. And when it stopped I could remember not one word that Derek had said. It was very strange. I was desperate to recall it because he told me all kinds of things that I knew I needed to know, I suppose, about what happened to him. But anyway, I was left with the peaceful feeling that he was in a wonderful place."

She told me that she met one other person who had the same experience; and the woman had said to Jane, "Whatever you do, if it happens again, don't try to get out of bed. When my husband came to me after his death, I, too, was desperate to write down what he said, and I got out of bed to get a pen. I had the same roaring sensation that you describe—and I was found in the morning in a coma."

I asked Jane what she thought was going on. "I don't know, but this woman told me that she believed that people who had died were on a different plane, and to hear them we entered that sphere. And that it takes time to return to Earth after such an experience.

"The strange thing is that after that experience I had with Derek, I had the strong feeling that if I really looked at the things Derek loved— the ocean, the storms, birds flying—and if I really *felt them*, then he would be able to share them, that somehow now he was in a different place—or 'plane,' as the woman said—he could only know things on Earth through a human's eyes. It was a very intense time."

Jane told me she did not usually talk about all of that—it was so strange, yet so real at the time.

"Jane, one last question. Why do you think it is that so many people say you give them hope?" I thought of my old college friend who had died by suicide and how many people were suffering and struggling with hopelessness.

"I honestly don't know—I wish I did. Perhaps it's because people realize that I am sincere. I unflinchingly lay out the grim facts—because people need to know. But then, when I lay out my reasons for hope, as I have in this book, they get the message and realize that there really could be something better if we get together in time. Once they realize that their life can make a difference, they have acquired a purpose. And, as we've said, having a purpose makes all the difference."

"I suppose it's time for us to close our conversation about hope and say goodbye—at least goodbye for now," I said. "Thank you, Jane; this exploration of hope has been wonderful."

"I always enjoy talking with you," she said. "I like to have my brain challenged."

"I've had my brain challenged, my heart opened, and my hope renewed," I replied.

"One moment," Jane said, taking the laptop over to the window. "There is one more being I want you to see, an old friend who's been with me ever since I came to the Birches when I was five years old. There— can you see him?"

And there he was—Beech, the tree that was deeded to her in the handwritten will Jane had her grandmother sign. Looking out into the garden at dusk, I could make out the black silhouette of Beech. I thought how fitting that we would close with a beech tree, considered to be the queen of British trees, having thrived since the last Ice Age.

"I know you can't see him properly in the dark," Jane said, "but

In Beech—one of my closest childhood friends. (JANE GOODALL INSTITUTE/COURTESY OF THE GOODALL FAMILY)

let me describe him to you. His bark is smooth and gray and his green leaves have recently turned to a soft autumn yellow and orange. And now they are beginning to fall.

"He's still standing," Jane added, "much taller than he was when I was a child. I couldn't climb him now, but I sit under him with a sandwich at lunchtime."

"Maybe someday when this pandemic is behind us I can join you for a sandwich under Beech," I said.

"We can always hope," said Jane.

"Well, I think that has to be the perfect quote to end our conversation on," I said.

After we waved goodbye and I closed my laptop, I thought of Jane on the other side of the world. For this day her work was over, but I knew that it would start again the next—Zooms and Skypes, taking her message of hope around a world that needed that message so desperately. "Good luck, Jane," I thought. And I felt another hope rise in me, that she would have the strength to continue for many more years. And I also knew that there would be a day when she would begin her next great adventure, binoculars and notebook at the ready. And that the indomitable human spirit in all of us would finish what she could not.

Conclusion:
A Message of Hope from Jane

This is my "studio" under the eaves of our family home, the Birches, where I have been "grounded" during the pandemic. It is also my bedroom. (RAY CLARK)

Dear Reader,

I am writing to you now from my home in Bournemouth on a very cold and very windy morning in February. It happens to be the start of the Lunar New Year, and I've been getting messages from all my Chinese friends—every one of them full of hope that it will be a better year than the last one. It was a year and a half ago that Doug and I began this conversation about hope in my home in Tanzania. And what a time it has been. First, Doug never got to Gombe because he had to rush back to America to be with his very sick father. Our second conversation worked out as planned—in the Netherlands. But the third, which was to have been here in Bournemouth so that Doug could see where I grew up, was first postponed and then canceled because of the pandemic. A pandemic that is still causing havoc around the world.

The tragedy is that a pandemic such as this one has long been predicted by those studying zoonotic diseases. Approximately 75 percent of all new human diseases come from our interactions with animals. COVID-19 is likely one of them. They start when a pathogen, such as a bacteria or virus, spills over from an animal to a human and bonds with a cell in a human. And this may lead to a new disease. Unfortunately for us, COVID-19 is highly contagious and it spread rapidly, soon affecting almost every country around the globe.

If only we had listened to the scientists studying zoonotic diseases who have long warned that such a pandemic was inevitable if we continued to disrespect nature and disrespect animals. But their warnings fell on deaf ears. We didn't listen and now we are paying a terrible price.

By destroying habitats we force animals into closer contact with people, thus creating situations for pathogens to form new human diseases. And as the human population grows, people and their livestock are penetrating ever deeper into remaining wilderness areas, wanting more space to expand their villages and to farm. And animals are hunted, killed, and eaten. They or their body parts are trafficked—along with their pathogens—around the world. They are sold in wildlife markets for food, clothing, medicine, or for the trade in exotic pets. Conditions in almost all of these markets are not only horribly cruel but usually extremely unhygienic—blood, urine, and feces from stressed animals all over the place. Perfect opportunity for a virus to hop onto a human—and it is thought that this pandemic, like SARS, was created in a Chinese wildlife market. HIV-1 and HIV-2 originated from chimpanzees sold for bushmeat in wildlife markets in Central Africa. Ebola possibly started from eating gorilla meat.

The horrific conditions in which billions of domestic animals are bred for food, milk, and eggs have also led to the spawning of new diseases such as the contagious swine flu that started on a factory farm in Mexico and noninfectious ones like *E. coli*, MRSA (staph), and salmonella. And don't forget that all the animals I've been talking about are individuals with personalities. Many—and especially pigs—are highly intelligent, and each one knows fear, misery, and feels pain.

But it is important to share the good, positive things that have emerged. During the various lockdown periods around the world, when there was less traffic and many industries were halted, fossil fuel emissions decreased significantly. Some people from the big cities had the luxury, perhaps for the first time, of breathing clean air and seeing the stars shine brightly in the night sky. Many people shared their delight in being able to hear the song of birds as the noise level decreased. Wild animals appeared in the streets of towns and cities. And although these things were temporary, it helped more people to understand what the world could—and should—be like.

Also, the pandemic has produced many heroes, like the doctors, nurses, and health care workers who risk—and too often lose—

their lives as they battle tirelessly to save others. A community spirit developed in many places—people helped each other. In one city in Italy people sang operatic arias back and forth between their balconies to boost their spirits. Brilliant television was produced. I especially loved it when a famous orchestra performed to an audience of plants—each one in its pot brought from the nearby botanical gardens and placed on one of the seats. And the crowning moment when the players rose and with great dignity and respect bowed to their horticultural audience. And the penguins from a zoo who were allowed to wander freely through an art gallery.

The human intellect was at work, too, developing new ways of connecting people virtually with each other. JGI held its first virtual global meeting—I did not think it could work, but although we missed the face-to-face, the fun, and hugging and just being together, things went smoothly—and we saved a lot of money. Today it is normal to hold conferences and business meetings by Zoom or one of the other incredible technologies. All of this is a great example of our adaptability and creativity.

Of course, it is desperately worrying for airlines and hotels; and in some countries, poaching of wild animals has increased because of the lack of tourists to support the hospitality industry, and the lack of funding for the salaries of the rangers who normally patrol the wildlife parks. It all points to the importance of using our creativity, our clever brains, and understanding and compassion, to create a more sustainable and ethical world in which everyone can make a decent living while existing in harmony with nature.

There are, in fact, many more people who have now realized this need for a new and more respectful relationship with animals and the natural world, and a new and more sustainable green economy. And there are signs that this is beginning to happen. Corporations are starting to think about the most ethical ways to source their materials, and consumers are thinking more carefully about their own ecological footprints. China has banned the eating of wild animals, and there is hope that the use of wild animal parts for medicine will also come to an end. Already the government has removed pangolin scales from the list of approved ingredients for traditional Chinese medicine. And there is a huge international effort to end the illegal trafficking of wild animals and plants. But, of course, we still have a long way to go.

What's more, there are many campaigns in several countries urging governments to phase out factory farms. Less and less meat is being consumed and more and more people are turning to a plant-based diet.

I have now been grounded since March last year, spending my days here in Bournemouth with my sister, Judy; her daughter Pip; and grandsons, Alex and Nickolai, twenty-two and twenty years old. And most of the time I am up in my little bedroom cum office cum studio under the caves. From where I had that last Zoom conversation with Doug.

At first, I was frustrated and angry. I felt terrible, having to cancel lectures and disappoint people. But I soon realized that I must face the inevitable and decided, with a small team of JGI staff, to create Virtual Jane. So many people have written to me, hoping that the enforced stay at home would be restful and give me time to meditate and generate new energy. In fact, as I told Doug, I have never, ever been busier or more exhausted in my life. Sending video messages around the

Pip, Judy, and me with Beech in the spring
garden of the Birches. (TOM GOZNEY)

world, taking part in conferences via Zoom or Skype or webinar or
some other technology, being interviewed, joining podcasts—and, in
fact, developing my own Hopecast!

Planning and giving virtual lectures is the hardest—somehow
you have to get the right energy into your presentation to inspire an
unseen audience when you have no feedback from an auditorium of
enthusiastic people. Instead, you are speaking to the tiny green light
of the laptop camera. And it's really hard, when you are talking to
people who are there on the screen and forcing yourself to look not
at them but at that little green light—so that from their viewpoint
you are looking at them!

Of course, I also terribly miss being with my friends, for when I was on the road, in between the lectures and press conferences and high-level meetings, there were the fun evenings getting together over Indian takeaway and red wine—and whisky of course! And having the opportunity to visit some amazing places and meet inspiring people. Instead, there are no breaks in the relentless schedule of Virtual Jane—just day after day gazing at a computer screen and talking into cyberspace.

But there is a bright side to all of this. I have been able to reach literally millions more people in many more parts of the world than I could possibly have done during my normal touring.

During my last Zoom conversation with Doug, I took him on a tour around my room and showed him many photos and other mementos from my travels. But there is so much more in almost every room of the house. I am surrounded by reminders of the various stages of my life. Here, in this much-loved house built in 1872, I am constantly reminded of my journey, the people, and the things that shaped me. Here are my roots that nurtured a shy, nature-loving child who has grown into a messenger of hope.

As I write to you on this cold, wet day in 2021, many countries have been hit by new, more contagious strains of the virus, all of which are hitching rides on unsuspecting human hosts and traveling around the globe, fueling yet more despair. Not surprisingly, therefore, a great deal of our attention is focused on bringing this pandemic under control.

But as a messenger, I have something very important I want to convey: we must not let this distract us from the far greater threat to our future—the climate crisis and the loss of biodiversity—for if we cannot solve these threats, then it will be the end of life on Earth as we know it, including our own. We cannot live on if the natural world dies.

During my lifetime we have defeated Nazism, although its fascist remnants are resurfacing. We defused the once great risk of nuclear Armageddon, even though these weapons still menace. And now we must defeat not only COVID-19 and its mutations—but also climate change and the loss of biodiversity.

It is somehow strange that my life has been sandwiched between world wars. The first, when I was a child, was a fight against human enemies, Hitler's Nazis. And now, as I approach my nineties, we must defeat two enemies, one against invisible, microscopic enemies; the other—our own stupidity, greed, and selfishness.

My message of hope is this: now that you have read the conversations in this little book, you realize that we can win these wars, that there is hope for our future—for the health of our planet, our societies, and our children. But only if we all get together and join forces. And I hope, too, that you understand the urgency of taking action, of each of us doing our bit. Please believe that, against all odds, we can win out, because if you don't believe that, you will lose hope, sink into apathy and despair—and do nothing.

We can get through the pandemic. Thanks to our *amazing human intellect* scientists have produced vaccines at record speed.

And if we get together and use our intellect and play our part, each one of us, we can find ways to slow down climate change and species extinction. Remember that as individuals we make a difference every day, and millions of our individual ethical choices in how we behave will move us toward a more sustainable world.

We should be so grateful for the incredible *resilience of nature*. And we can help the environment heal not only by means of the big restoration projects but as a result of our own efforts as we choose how to live our lives and think about our own environmental footsteps.

There is great hope for the future in the actions, the determination and energy of *young people* around the world. And we can all do our best to encourage and support them as they stand up against climate change and social and environmental injustice.

Finally, remember that we have been gifted not only with a clever brain and well-developed capacity for love and compassion, but also with an *indomitable spirit*. We all have this fighting spirit—only some people don't realize it. We can try to nurture it, give it a chance to spread its wings and fly out into the world giving other people hope and courage.

It's no good denying that there are problems. It is no shame if you think about the harm we've inflicted on the world. But if you concentrate on doing the things you *can* do, and doing them well, it will make all the difference.

During one of my visits to Tanzania, where Roots & Shoots began, I attended an event where all the groups of the neighborhood came together to share their projects and socialize. There was a lot of laughter and a lot of enthusiasm.

As the event wrapped up, everyone there got together and shouted out, "Together we can"—meaning together they could set the world right. I took the mike and told them, "Yes, absolutely we *can*. But *will* we?" This startled them, but they thought about it and understood what I meant. I led them in a rousing "Together we *can*. Together we *will!*" This is now the way they end all their meetings, and it has spread to other countries. And I sometimes end my lectures that way. I gave a short talk at the second largest music festival in Europe—to a crowd of around sixteen thousand people. I asked them to join me in that clarion call for action. There was a response, but it was not impressive. I told them, primary school children did better than that, and we tried again. I still get goose bumps when I remember how the entire audience rose to their feet, and the words rang out in the warm evening air.

But when the same scenario was repeated at Davos at the beginning of last year, when I gave a talk to powerful CEOs of big corporations with a smattering of politicians and other attendees—that was more than amazing. Their first response, again, was weak. But when I told them I had hoped for more enthusiasm to show their commitment to change, and when they all stood and gave a loud and ringing response, followed by prolonged applause, there were tears in my eyes.

Together we CAN! Together we WILL!

Yes, we can, and we will—for we must. Let us use the gift of our lives to make this a better world. For the sake of our children and theirs. For the sake of those struggling in poverty. For the sake of the lonely. And for the sake of our brothers and sisters in the natural world—the animals, the plants, the trees.

Please, please rise to the challenge, inspire and help those around you, play your part. Find your reasons for hope and let them guide you onward.

Thank you,

Jane Goodall

Acknowledgments

From Jane:

After eighty-seven years how can I properly acknowledge all the people who have helped me on my way, kept me going when times were hard, encouraged me to do things I thought I could not do?

Of course, I must start with my wonderful mother and the rest of my family. Their role is well described in this book. Rusty, who taught me that we are part of the animal kingdom. Louis Leakey, who gave me the opportunity to realize my dreams, who had faith in a young girl who went into the field with only her passion for learning about chimpanzee behavior. Leighton Wilkie, who provided money for my first six months in the field. David Greybeard, who allowed me to watch him using and making tools—an observation that so interested the National Geographic Society that they continued to fund my research. BIG THANKS. And I owe so much to my first husband, Hugo van Lawick, whose film and stills enabled me to persuade the animal behavior people of the time that we are not the only beings with personalities, minds, and emotions.

There are so many people and animals who have contributed to my understanding of the world around us, who have helped me during my life's journey. They are too many to mention. The students and scientists who came to Gombe and have enriched our understanding of chimpanzee and baboon behavior. I single out Dr. Anthony Collins because he has been with me since 1972, has helped to keep Gombe going, and is always there to help and support me in my travel in Tanzania, Burundi, Uganda, and DRC. My second husband, Derek Bryceson, played a major role in securing the continuation of the work at Gombe. His relationship with the Tanzanian government allowed us to get briefly to Gombe when it was closed off after the students were kidnapped—we were taken there in a military helicopter. And how wonderful that the field assistants continued to follow the chimpanzees and baboons even when, for a while, I was not able to stay for more than a few days.

My heartfelt gratitude to the staff and volunteers of the Jane Goodall Institutes and our Africa programs in Tanzania, Uganda, DRC, Republic of Congo, Burundi, Senegal, Guinea, and Mali. And to those working to improve the welfare of animals in zoos and especially in our Tchimpounga and Chimp Eden sanctuaries for orphan chimpanzees, and the other sanctuaries that I helped to establish—Ngamba Island, Sweetwaters, and Tacugama.

Then there is a group of people who have supported me during the pandemic and made it possible for me to continue reaching out to people around the world through technology: Dan DuPont, Lilian Pintea, Bill Wallauer, Shawn Sweeney, Ray Clark, and the hardworking team in GOOF—the Global Office of the Founder—Mary Lewis, Susana Name, and Chris Hildreth. I am so grateful to Carol Irwin for her wise council throughout many challenging times. Grateful thanks to Mary Paris, who is the guardian of a large archive of my photos and whose patience and magical skills made it possible for us to include all the photos in this book. And a very special thank-you to all the young—and not so young—people organizing and taking action in our Roots & Shoots programs all around the world. For it is this movement that gives me so much hope for our future.

Finally, I come to those who have helped to make this book possible. All of you who have contributed stories and photographs, far too many to mention by name. The last in-person conversations Doug and I had were in the Netherlands, and we were so really grateful to Patrick and Daniëlle van Veen, who found that wonderful forester's hut in the wood, provided food and wine, and Daniëlle cooked delicious food. Thank you so much.

And to those doing the actual work: including the wonderful team at Celadon Books, especially assistant editor Cecily van Buren-Freedman and most especially our wonderful and supportive editor Jamie Raab, president and publisher of Celadon Books, who has guided this book with such care and attentiveness while having to endure many delays because of my terrible schedule. And my endless gratitude to Gail Hudson, with whom I have often collaborated in the past, and who has been such a support to me as I struggled to write at the same time as dealing with everything else. Thank you, Gail. I would be remiss if I did not pay tribute to my sister, Judy Waters, and her daughter, Pip, who have kept me going through these tough days, doing the shopping and cooking so that I could devote full time to working. I am so grateful to Adrian Sington, who encouraged me to collaborate with Doug Abrams on a book of hope. And last, of course, Doug himself. He conceived of doing this book in the first place and with his penetrating questions managed to draw out some of my inmost thoughts. And he patiently adjusted his schedule to fit in with my increasingly crazy one during our final Zoom discussions about the meaning of and reasons for hope.

From Doug:

As I learned while writing this book, hope is a social gift, one that is nurtured and sustained by those around us. Each of us has a web of hope that supports, encourages, and uplifts us throughout our lives. I have been blessed by so many people who have helped me in countless ways.

First, I must thank my mother, Patricia Abrams, and my late father, Richard Abrams, who believed in me even when I did not believe in myself. Also, my brother, Joe, and my sister, Karen, who have been lifelong friends as well as siblings.

My extended family, teachers, friends, and colleagues have also been there all along the journey of life and particularly in creating this book as my father died and my son struggled with his brain injury. In particular, I'd like to thank my amazing friends Don Kendall, Rudy Lohmeyer, Mark Nicolson, Gordon Wheeler, Charlie Bloom, Richard Sonnenblick, Ben Saltzman, Matt Chapman, and Diana Chapman. I'd also like to thank my brilliant and fun friends and colleagues at Idea Architects who helped conceive, envision, and create this book, including Boo Prince, Cody Love, Staci Bruce, Mariah Sanford, Jordan Jacks, Stacie Sheftel, and especially the brilliant Esmé Schwall Weigand, who worked tirelessly to help with research and editing all along the way, and Lara Love Hardin and Rachel Neumann, my constant guides through the literary forest and partners in creating a wiser, healthier, and more just agency and world. Boo and Cody could not have been a better production team and traveling companions to Tanzania and could not have been more understanding when

I needed to leave in the middle of our trip when my father was admitted to the hospital. I also want to thank our incredible foreign rights team, Camilla Ferrier, Jemma McDonagh, and Brittany Poulin of the Marsh Agency, and Caspian Dennis and Sandy Violette of Abner Stein, who helped share the book with the world. This project would not exist without the love and matchmaking of my beloved friends and authors Christiana Figueres and Tom Carnac, two of the architects of the Paris Climate Agreement and two of the people whom history will remember as giving humanity a fighting chance. They introduced me to Jane and encouraged the project all along the way.

I would not last long without the love and support of my brilliant wife, Rachel, and our children, Jesse, Kayla, and Eliana, who are three of my greatest hopes for the future and demonstrate the power of young people, each in their own unique way.

As Jane has said, the entire team at Celadon was incredible to work with and saw the vision and potential of this book from the beginning, including Cecily van Buren-Freedman, Christine Mykityshyn, Anna Belle Hindenlang, Rachel Chou, Don Weisberg, Deb Futter, and most of all, Jamie Raab. Jamie has been someone I have long admired as one of the most brilliant and creative publishers in the world, and she has been a joy to work with from start to finish, helping to guide the project with her wisdom, kindness, and deep knowledge of the reader's hopes and dreams.

I'd like to thank all at the Jane Goodall Institute who helped with this project, from my first conversations with Susana Name to my joyous lunch with Mary Lewis, who has been there at every step of the way with her warmth, her insight, and her ability to make miracles happen in Jane's impossibly crowded schedule. Adrian Sington, Jane's literary agent, has been a catalyst and a cherished colleague who made the project possible even against so many odds and in the face of the global pandemic. Our first meeting at the London Book Fair is one of the very happy memories of my life. Gail Hudson, Jane's longtime collaborator and friend, helped us immensely as we wove our dialogues together. She was instrumental to the book's completion and has become my trusted friend and advisor, too.

And finally, I would like to thank Jane for the great gift of self that she has given the world in this book. I sought out Jane because she is a naturalist who has rare and necessary knowledge of our world, but I also discovered a humanitarian and wisdom figure who speaks for us and for the Earth. As a poet and writer, her devotion to making sure that every word expressed her greatest truth was deeply inspiring. It has been one of the great privileges of my life to companion Jane into the depths of her understanding of human nature and how hope might be part of what saves us. Despite the extraordinary demands of an aching world desperate for her guidance, she was extremely generous with her time, her wisdom, and her friendship, first as I was traveling the rocky terrain of personal grief and then during the unprecedented global pandemic that revealed to all of us how vulnerable and precious our world truly is.

Further Reading

I: What Is Hope?

For a deeper exploration of Jane's life and the experiences that have formed her views, see her spiritual autobiography, *Reason for Hope: A Spiritual Journey* (Warner Books, 1999). For more information about her work with chimpanzees, see her classic works on the chimpanzees of Gombe, *In the Shadow of Man* (Houghton Mifflin, 1971) and *Through a Window: My Thirty Years with the Chimpanzees of Gombe* (Houghton Mifflin, 1990).

For more on the subject of hope studies, see Charles Snyder's *Psychology of Hope: You Can Get There from Here* (Free Press, 1994); Shane Lopez's *Making Hope Happen: Create the Future You Want for Yourself and Others* (Atria Paperback, 2014); and Casey Gwinn and Chan Hellman's *Hope Rising: How the Science of HOPE Can Change Your Life* (Morgan James, 2019). There's also an excellent brief article by Kirsten Weir, writing for the American Psychological Association ("Mission Impossible," *Monitor on Psychology* 44, no. 9 [October 2013], www.apa.org/monitor/2013/10/mission-impossible).

The idea that when we think of the future we're either fantasizing, dwelling, or hoping comes from Lopez's book cited above (p. 16) as does the meta-analysis of hope's impact on academic success, workplace productivity, and overall happiness (p. 50).

In another study, psychologists at the University of Leicester looked at students over three years and found that the more hopeful students did better academically. In fact, hope mattered more than intelligence, personality, and even prior academic achievement. ("Hope Uniquely Predicts Objective Academic Achievement Above Intelligence, Personality, and Previous Academic Achievement," *Journal of Research in Personality*, 44 [August 2010]: 550–53, https://doi.org/10.1016/j.jrp.2010.05.009). In another study, researchers compared the relationship between hope and productivity in their analysis of forty-five studies that examined more than eleven thousand employees in a variety of fields. ("Having the Will and Finding the Way: A Review and Meta-Analysis of Hope at Work," *Journal of Positive Psychology* 8, no. 4 [May 2013]: 292–304, https://doi.org/10.1080/17439760.2013.800903). They concluded that hope determines 14 percent of workplace productivity, which was more than other measures including intelligence or optimism.

Hope can impact us collectively as well as individually. In a survey of a thousand people in a medium-sized city, researcher Chan Hellman found that collective hope was the most significant predictor for overall community well-being. When the survey was connected to public health data, they even found that both individual hope and collective hope predicted life expectancy (Hellman, C. M., & Schaefer, S. M. [2017]. *How hopeful is Tulsa: A community wide assessment of hope and well-being.* Unpublished manuscript.)

Other research shows that hope seems to impact our physical health. Stephen Stern, a physician at the University of Texas

Health Science Center at San Antonio, and his colleagues conducted a mortality study of almost eight hundred Mexican Americans and European Americans (Stephen L. Stern, Rahul Dhanda, and Helen P. Hazuda, "Hopelessness Predicts Mortality in Older Mexican and European Americans," *Psychosomatic Medicine* 63, no. 3 [May-June 2001]: 344–51, doi: 10.1097/00006842–200105000–0 0003). When controlling for gender, education, ethnicity, blood pressure, body mass index, and drinking behavior, the people who were less hopeful were more than twice as likely to have died from cancer and heart disease within three years. Stern believes that hope for the future drives our behaviors in the present, and the choices we make in the present determine whether we have a longer or shorter life.

The components of the hope cycle originate with Charles Snyder who identified them in his book, *Psychology of Hope* (Simon & Schuster, 2010), as goals, willpower (often called agency or confidence), and waypower (often called pathways or realistic ways to realize one's goals). Other researchers, including Kaye Herth, who developed the Herth Hope Index, have included social support as one of the building blocks of hope ("Abbreviated Instrument to Measure Hope: Development and Psychometric Evaluation," *Journal of Advanced Nursing* 17, no. 10 [October 1992: 1251–59, doi: 10.1111/j.1365–2648.1992.tb01843.x).

For more about Edith Eger, see her books *The Choice: Embrace the Possible* (Scribner, 2017) and *The Gift: 12 Lessons to Save Your Life* (Scribner, 2020).

Reason 1: The Amazing Human Intellect

For an explanation of the neuroscience of hope and optimism, see Tali Sharot's *The Optimism Bias: A Tour of the Irrationally Positive Brain* (Pantheon, 2011). As Sharot points out, the frontal cortex, which is larger in humans than in other primates and likely the neural basis for the human intellect that Jane references, is essential for language and goal setting, and likely also for hope and optimism. Sharot identified a specific part of the frontal cortex, the rostral anterior cingulate cortex (rACC), that influences emotion and motivation and may contribute to hope. In her research, the more optimistic a person was, the more likely they were to imagine positive future events with great vividness and detail. As subjects thought of positive events, this part of the brain was activated more and seemed to be connecting and modulating the amygdala, an ancient structure in the brain associated with emotion, especially fear and excitement. The rACC in optimistic people seems to calm the fear that is aroused when they imagine negative events and gets them more excited when they think of positive events. This may be the neural basis for Lopez's felicitous phrase that humans are hope-fear hybrids (Lopez, p. 112).

For more on the intelligence and communication of trees, see Suzanne Simard's *Finding the Mother Tree: Discovering the Wisdom of the Forest* (Alfred A. Knopf, 2021) and Peter Wohlleben's *The Hidden Life of Trees: What They Feel, How They Communicate—Discoveries from a Secret World* (Greystone Books, 2016).

Reason 2: The Resilience of Nature

For more stories about the resilience of nature and more detail on some of the stories Jane told me, see Jane's books *Hope for Animals and Their World: How Endangered Species Are Being Rescued from the Brink* (Grand Central Publishing, 2009) and *Seeds of Hope: Wisdom and Wonder from the World of Plants* (Grand Central Publishing, 2014).

For more information on the dramatic loss of biodiversity and rapid extinction, see the May 2019 UN report, "Nature's Dangerous Decline 'Unprecedented'; Species Extinction Rates 'Accelerating,'" Sustainable Development Goals, www.un.org/sustainabledevelopment /blog/2019/05/nature-decline-unprecedented-report/.

For the APA report on the effects of climate change on mental health, see Susan Clayton Whitmore-Williams, Christie Manning, Kirra Krygsman, et al., "Mental Health and Our Changing Climate: Impacts, Implications, and Guidance," March 2017, www.apa.org /news/press/releases/2017/03/mental-health-climate.pdf.

For more on the ability of ecosystems to recover, see the study "Rapid Recovery of Damaged Ecosystems" by Holly P. Jones and Oswald J. Schmitz of the Yale University School of Forestry and Environmental Sciences (*PLOS ONE*, May 27, 2009, https://doi.org/10.1371/journal .pone.0005653). After reviewing 240 independent studies spanning a hundred years, they found that ecosystems could recover when the source of the pollution and destruction is stopped. The ecosystems they studied recovered within a decade to a half century, with forests recovering on average in forty-two years and ocean floors recovering on average in ten years. Environments with multiple sources of destruction took an average of fifty-six years to recover, but some ecosystems were pushed beyond the point of return and never recovered, although even these might recover in much larger time scales that may not be relevant to human civilization. The researchers qualified their findings by saying that even these heavily damaged ecosystems can recover "given human will."

For more on our need for nature and the deep influence that nature has on human health and well-being, see Caoimhe Twohig-Bennett and Andy Jones's "The Health Benefits of the Great Outdoors: A Systematic Review and Meta-Analysis of Greenspace Exposure and Health Outcomes," in which researchers analyzed over 140 studies involving more than 290 million people in twenty countries and found that spending time in nature or living close to nature resulted in diverse and significant benefits including a reduction in type II diabetes, cardiovascular disease, premature death, and preterm birth (*Environmental Research* 166 [October 2018]: 628–37, doi: 10.106/j .envres.2018.06.030). While the reason nature has such a profound impact is not clear, one theory is that nature seems to reduce participants' stress as measured by their salivary cortisol level.

Environmental neuroscientist Marc Berman at the University of Chicago and his colleagues have found that having more trees on a street is related to the improved health of its residents (Omid Kardan, Peter Gozdyra, Bratislav Misic, et al., "Neighborhood Greenspace and Health in a Large Urban Center, *Scientific Reports* 5, 11610 [July 9, 2015], https://doi.org/10.1038/srep11610). People who lived on a street with ten more trees had health improvements that were related to being seven years younger than those who lived on streets with less tree cover, even when controlling for other confounding factors such as income and education. Berman doesn't yet know why this is, but he suspects it might have to do with air quality, and also with the soothing aesthetic that nature provides. In another study, he found that a simple walk in nature leads to 20 percent increases in working memory and attention and that people can also experience cognitive benefits from images, sounds, and videos of nature. (Marc G. Berman, John Jonides, and Stephen Kaplan, "The Cognitive Benefits of Interacting with Nature, *Psychological Science* 19, no. 12 [December 2008]: 1207–12, https://doi.org/10.1111/j.1467–9280.2008.02225.x; Marc G. Berman, Ethan Kross, Katherine M. Krpan, et al., "Interacting with Nature Improves Cognition and Affect for Individuals with Depression," *Journal of Affective Disorders* 140, no. 3 [November 2012]: 300–305, https://doi.org/10.1016/j.jad.2012.03.012.

For more on the Davos World Economic Tree Planting Initiative, see "A Platform for the Trillion Tree Community," www.1t.org/. The research that demonstrated "The Global Tree Restoration Potential" that led to the initiative was published in *Science* by Thomas Crowther et al. (365, no. 6448 [July 5, 2019]: 76–79, https://science.sciencemag.org/content/365/6448/76).

Reason 3: The Power of Young People

For more on Roots & Shoots programs, see http://rootsandshoots.org/.

Chan Hellman's story was relayed to me in a phone interview, but an account can be found in his book *Hope Rising* (Morgan James, 2019).

Reason 4: The Indomitable Human Spirit

For a wonderful video on Jia Haixia and Jia Wenqi and their tree-planting friendship, see *GoPro: A Blind Man and His Armless Friend Plant*

a Forest in China (www.youtube.com/watch?v=Mx6hBgNNacE&t
=2s) and learn more at https://gopro.com/en/us/goproforacause
/brothers.

Becoming a Messenger of Hope

For more on near-death experiences and what they may say about
life after death, see Elisabeth Kübler-Ross's classic *On Life After Death*
(Celestial Arts, 2008) or Bruce Greyson's more recent book, *After:
A Doctor Explores What Near-Death Experiences Reveal About Life and
Beyond* (St. Martin's Essentials, 2021). Greyson, who is the leader of
the field of near-death studies, has been examining these experiences
for over forty years. He studied many people who, while near death,
had seen and learned things that should not have been possible, like
meeting relatives that they did not know they had. He said that after
people have near-death experiences they almost universally believed
that death is not something to fear and that life or consciousness
continues in some form beyond the grave. Near-death experiences
also transform how people live their lives and inspire a belief that
there is meaning and purpose in the universe. Some of the most
fascinating stories relate to what Jane said about this life possibly
being a test. According to Greyson's research, many people experience an end-of-life review where they literally do see their whole life
flash before them and understand why certain events in their lives
happened.

For more on Francis Collins's views, see his book, *The Language of
God: A Scientist Presents Evidence for Belief* (Free Press, 2006).

For other books in the Global Icons Series, go to www.ideaarchitects .com/global-icons-series/.

For further information about Jane Goodall's work, see www .janegoodall.global and www.rootsandshoots.global.